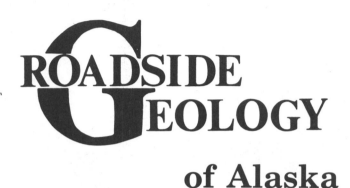

ROADSIDE GEOLOGY

of Alaska

Cathy Connor
Daniel O'Haire

MOUNTAIN PRESS PUBLISHING COMPANY
MISSOULA 1988

Library of Congress Cataloging-in-Publication Data

Connor, Cathy L.
Roadside geology of Alaska.

(Roadside geology series)
Bibliography: p.
Includes index.
1. Geology—alaska—Guide-books   2. Alaska—
Description and travel—1981-   —Guide-books.
I. O'Haire, Daniel.   II. Title   III. Series.
QE83.C74   1988   557.98   88-1651
ISBN 0-87842-213-7 (pbk.)

Printed in the U.S.A.

Mountain Press Publishing Company
P.O. Box 2399
Missoula, MT 59805
(406) 728-1900

To the men and women who have dedicated their lives to studying Alaska's geology . . . and to my grandparents Romer and Leo Frantzen who helped me all along the way.

—Cathy Connor

Thanks to my *Roadside Geology* companions Steve, John, and Tom, and to our editors Don and Dave.

–Dan O'Haire

Preface

This book is designed for the amateur, as well as the professional geologist. We have tried to avoid the use of technical terms and have attempted to explain the geology of Alaska using language that anyone can understand.

Roadside Geology takes an inquisitive, philosophical approach to the subject. The focus is often on the big statewide picture rather than specific details and if certain local items are overlooked, it is because we had so much territory to cover.

Nearly all of the information in this book was first gathered elsewhere by our professional colleagues. We plowed through many volumes of geologic literature to glean the information which we present here for the reader. We apologize for failing to credit our colleagues for their scientific contributions but crediting each and every geologist who contributed to the understanding of the geology of our state would have added an additional volume to this book.

Readers familiar with the Roadside Geology series will recognize that *Alaska* is different. We do not cover the entire state, but only those portions which have an established transportation network and at least a few residents. Roads that go to a single destination and dead-end are discussed starting with the access point. Elsewhere, the routes are described east to west or south to north. The ferry routes are treated just like roads because the marine highway system is an extension of the roadways in Alaska.

We divided up the state into two portions. Cathy Connor lives in Juneau and was responsible for southeast Alaska, the eastern Alcan Highway, and the Copper River basin. Dan O'Haire lives in Anchorage and was responsible for south-central and interior Alaska, plus the Pipeline Haul Road and Nome. We hope our collaboration meshes neatly into a cohesive whole.

Some of the material in this book is from unpublished reports of the United States Geological Survey, including some of the names and descriptions of terranes. Some of the material is our own contribution and appears in print here for the first time. We are indebted to our Roadside Geology editors, David Alt and Donald Hyndman of the University of Montana.

The Authors

Contents

GEOLOGIC TIME SCALE

After Geological Society of America 1983

PRECAMBRIAN

ALASKAN ROCK RECORD

Alexander terrane forms far from southeast Alaska.

Sediments that will become the Yukon-Tanana Terrane form in Canada.

EON: PROTEROZOIC / ARCHEAN

AGE: 750 — 1000 — 1250 — 1500 — 1750 — 2000 — 2250 — 2500 — 2750 — 3000 — 3250 — 3500 — 3750

PALEOZOIC

ALASKAN ROCK RECORD

Fusilinids live in Tethys Sea. They will later provide clues to Alaska's history.

Wrangellia terrane forms far from Alaska.

World wide tropical climate. Brooks Range rocks form in northern Canada.

No Alaska as we know it today.

Trilobites live that will one day occur as fossils along the Yukon River near Eagle, Alaska.

PERIOD: PERMIAN / CARBONIFEROUS (PENNSYLVANIAN, MISSISSIPPIAN) / DEVONIAN / ORDOVICIAN / CAMBRIAN

AGE: 260 — 280 — 300 — 320 — 340 — 360 — 380 — 400 — 420 — 440 — 460 — 480 — 500 — 520 — 540 — 560

MESOZOIC

ALASKAN ROCK RECORD

Duck-billed dinosaurs live in Northern Alaska.

Huge eruptions of lava in Copper River Basin.

Basins form in Arctic Ocean that will trap Prudhoe Bay oil.

Terranes are added onto N. America that will become Alaska.

Break up of Pangaea. North America drifts westward over Pacific.

PERIOD: CRETACEOUS / JURASSIC / TRIASSIC

AGE: 70 — 80 — 90 — 100 — 110 — 120 — 130 — 140 — 150 — 160 — 170 — 180 — 190 — 200 — 210 — 220 — 230 — 240

CENOZOIC

ALASKAN ROCK RECORD

Ice Age

Aleutian Islands Born

Ferns grow that will become Kenai Penin. fossils.

Subduction of Kula Plate. Prince Wm. Terrane docks.

Oil bearing sediments form in Cook Inlet.

EPOCH: HOLOCENE / PLEISTOCENE / PLIOCENE / MIOCENE / OLIGOCENE / EOCENE / PALEOCENE

PERIOD: QUATERNARY / NEOGENE / PALEOGENE / TERTIARY

AGE: 5 — 10 — 15 — 20 — 25 — 30 — 35 — 40 — 45 — 50 — 55 — 60 — 65

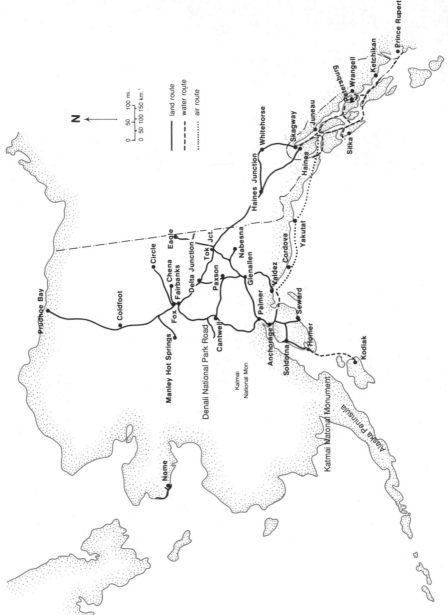

Travel routes covered in this book.

INTRODUCTION

The rocks that make up Alaska began their journey there, like most of the state's residents, from points far away. Large pieces of the earth's crust, perhaps as many as fifty, drifted into place and became geologically glued together to form the state of Alaska. The state has grown like a snowman by adding material to its edges. The pieces have been added in waves, the last one beginning about 200 million years ago and continuing through to the present.

The entire Pacific Coast of the North American continent, from Baja California to the end of the Aleutian Chain in western Alaska, has been transplanted and grafted on piecemeal. Alaska is at the northern receiving end of a conveyor belt of ready-made blocks of crust, some of which have traveled thousands of miles across the Pacific basin. These blocks have been as large as several thousand square miles. In some cases they have survived their journey with relatively little wear and tear.

For a long time an important section of the conveyor belt was the Kula plate, a now vanished oceanic plate that was pushed northward by the Pacific Plate. About 40 million years ago, the Kula plate slid down an oceanic trench; it was subducted beneath North America. Today the sole surviving piece of the Kula plate lies beneath the shallow water of the Bering Sea.

Some of Alaska's blocks began as ocean crust formed by the eruption of undersea volcanoes and vents along giant sea-floor rifts like those near the Galapagos Islands today. Other blocks

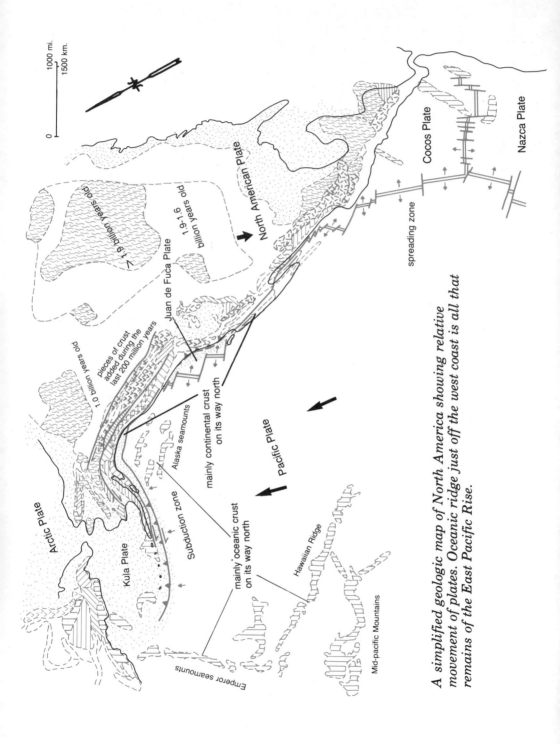

A simplified geologic map of North America showing relative movement of plates. Oceanic ridge just off the west coast is all that remains of the East Pacific Rise.

1000 mi.
1500 km.
0

North American Plate

1.9 billion years old

1.9-1.6 billion years old

Juan de Fuca Plate

pieces of crust added during the last 200 million years

1.0 billion years old

mainly continental crust on its way north

Alaska seamounts

Arctic Plate

Subduction zone

Kula Plate

Pacific Plate

mainly oceanic crust on its way north

Hawaiian Ridge

Emperor seamounts

Mid-pacific Mountains

spreading zone

Cocos Plate

Nazca Plate

were created as chains of volcanic islands formed in areas where moving slabs of denser oceanic crust sank into the earth's interior and pushed beneath masses of continental crust. Still other blocks formed by accumulation of sediments eroded from the continents.

As the blocks moved across the Pacific Ocean and made contact with the North America continent, they were often sliced up along faults into narrow strips parallel to the continental margin. The Yakutat block, now arriving in Alaska's southern coast, is sliding along the Queen Charlotte-Fairweather fault system, sandwiched between the Pacific and North American plates.

The continuing rise of the St. Elias Mountains, earthquakes along the coast, and rerouting of many rivers in the Yukon, are just some of the effects of this newly arrived block merging into southern Alaska.

Plate Tectonics

The earth's land masses ride on large pieces of lithosphere, the earth's crust and more rigid part of the upper mantle, that are in continuous motion relative to one another. Beneath the lithosphere, heat causes convection currents to flow slowly in the plastic deeper part of the mantle. At places where convection currents rise through the mantle, rift zones form on the ocean floor; magma wells up to fill the sea floor fractures and erupts on the sea floor forming new oceanic crust. Crustal plates move along on the convection currents. At the edge of ocean basins, plates collide with the thick, less dense crust of adjacent continents. There the oceanic crust and upper mantle sink beneath the more buoyant continental crust along deep-sea trenches. The size of the earth is not changing so an equal amount of oceanic crust must be destroyed for all that is made.

The southern coast of Alaska is an active zone of plate collision, an area where the continent is growing by the addition of seaboard terranes. Here, oceanic crustal rocks are being forced beneath the continent. Deep-sea sediments and some volcanic rocks are too buoyant to be dragged down a trench so the less dense material is scraped off onto the continental margin. Many of the rocks along Alaska's southern margin were formed as such trench wall scrapings. Geologists call them melanges.

The subducted material is heated up and partially melted as it moves beneath the continent, generating magma that rises and erupts to form volcanoes at the surface.

The Making of Alaska

Unravelling Alaska's complex and fragmented geologic history would have baffled Sherlock Holmes. Deciphering Alaska's geologic mysteries has taken many summers of geologic field work and the job is far from over. Progress has been slowed by the ruggedness and remoteness of the land. The geologic maps of Alaska are now in their early stages compared to those available in the lower 48 states. Important geologic clues in Alaska are covered by ice caps, by great expanses of tundra or alder bushes. Large brown bears who dislike disturbances caused by rock hammering, outboard motors, or helicopters, guard some of the clues.

Fossil Evidence of Terrane Movement

The fossil record contains important evidence of earth history. Marine organisms now preserved as fossils in the far north spent their lives in tropical waters. The present distribution of these fossils provides information that enables geologists to reconstruct plate movements. These animals lived in the Tethys Sea, which during Permian time separated Europe and Asia to the north from Africa, India, Tibet, and Australia to the south. The presence of 'Tethyan' fossils in Alaska and British Columbia suggests that the terranes on which they are riding moved across the Pacific Ocean basin from the original Tethys Sea.

Fossil Magnetic Fields

For unknown reasons the earth's magnetic field sometimes reverses itself so that the north magnetic pole becomes the south magnetic pole. The reversals are preserved in the rock record in the orientation of magnetite crystals that align themselves as they cool in volcanic lavas, or in some iron-bearing sedimentary rocks. Both the direction of magnetic north and the inclination of the magnetic field at the latitude of formation

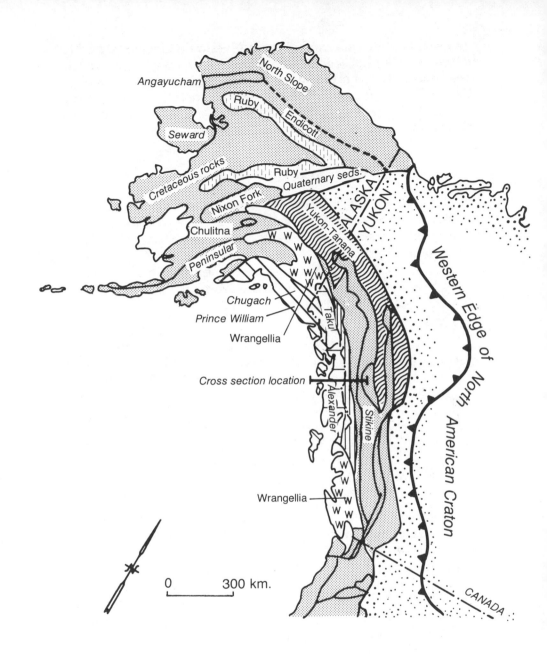

The mosaic of pieces in western North America.

are preserved by this iron mineral. This information is used to reconstruct the latitude at which the rock formed and how the rock mass has rotated since it formed. Terranes such as Wrangellia originated either 15 degrees north or south of the equator, either way a considerable distance from modern Alaska latitudes of more than 54 degrees north.

Late Precambrian and Paleozoic

Before about 200 million years ago, the area we know as Alaska did not exist. The western coast of North America ended about 200 km (120 miles) farther inland than now, along a line roughly parallel to the western edge of the Rocky Mountains. The history of Alaskan geology begins with the addition of the Yukon-Tanana Terrane to the North Amerian continent about 200 million years ago.

The Birch Creek Schist is the oldest rock in interior Alaska and makes up most of the Yukon-Tanana terrane. It consists of metamorphic rocks; muscovite-quartz schist, micaceous quartzite, and lesser amounts of graphitic schist. It has been so complexly folded that estimates of its true thickness are nearly impossible. It extends from an area west of Fairbanks east to Dawson in the Yukon Territory. The schist is thought to be late Precambrian or earliest Paleozoic in age, between 600 to 800 million years old. It formed through metamorphism of shale, mudstone, and sandstone originally deposited along the western margin of North America.

In northern Alaska along the southwestern Brooks Range, a schist belt shows evidence of having undergone a late Precambrian metamorphism that created greenschist and blueschist, uncommon rocks. Blueschist metamorphism is common along active continental margins where oceanic rocks slip under continental margins; it forms under low temperature high pressure conditions. The schist may underlie much of the Brooks Range and appears in deep wells farther to the north, in Prudhoe Bay.

On the Seward Peninsula in the Kigluaik Mountains northeast of Nome, gneisses produced by the metamorphism of sedimentary rocks have also been dated as late Precambrian. In southeastern Alaska, metamorphosed sedimentary and volcanic rocks of Precambrian age are found on Prince of Wales Island.

Mid-Paleozoic time was a period of changing tectonic conditions as a marine basin deepened and was affected by volcanism covering at least 5,000 square kilometers. Sandstone, mudstone, and carbonate were deposited on a shelf edge of the North American continent. The shelf was broad and gently sloping and had a tropical or subtropical climate, conditions that exist today on the Bahamas Bank in the Carribean. Sedimentary rocks destined to become parts of Alaska were deposited in the belt stretching from Ellesmere Island in northern Canada to California.

From late Cambrian to early Devonian time, between about 370 to 515 million years ago, the North American plate shifted. This disturbance caused volcanoes to erupt and sediments to be eroded and redeposited. The results exist today in the Keevy Peak Formation in the Northern Alaska Range. Volcanic rocks, later metamorphosed, are now the Totatlanika Schists in interior Alaska and rocks of the Alexander terrane in southeastern Alaska. These terranes were later sliced, transported north, and then assembled in Alaska, along with terranes that may have originated in places as diverse and distant as Asia, the South Pacific, or South America.

Mesozoic Alaska

All of today's continents joined together by late Paleozoic time to form the supercontinent known as Pangaea, but it didn't last long. About 180 million years ago, this giant continental mass began to break up. North America and South America moved west away from Europe and Africa, as a newly formed oceanic ridge in the Atlantic forced the continents apart. As North and South America moved west, they pushed against the oceanic crust of the Pacific. The oceanic plates, moving east or north from oceanic ridges in the Pacific basin, began to sink through trenches along the margins of the American continents.

As a result of this subduction, a string of volcanoes erupted massive quantities of lava along the west coast of North America, laying the foundations for the modern mountain belt. The intensity of mountain building declined in late Cretaceous time about 80 million years ago. Erosion then dominated for a time and wore the mountains down. The remains of these mountains, their roots, and the sediments eroded from them would later become part of Alaska.

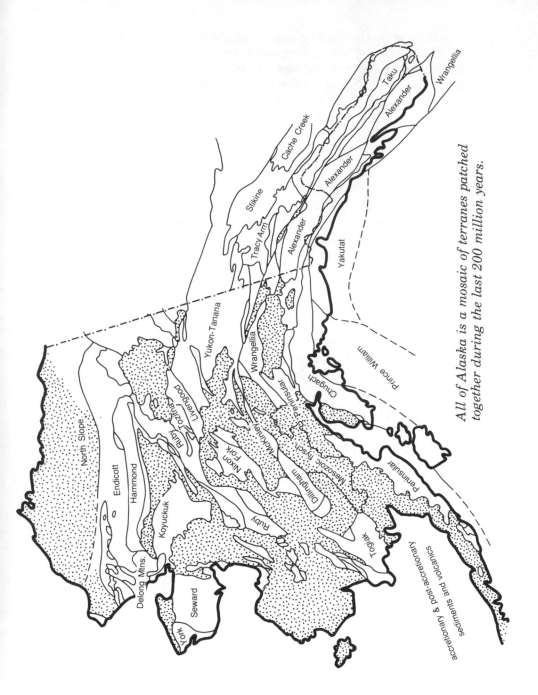

All of Alaska is a mosaic of terranes patched together during the last 200 million years.

Southern Alaska and its Terranes

In Middle Triassic time, the Wrangellia, Alexander, and Stikine terranes were somewhere offshore of the coast of North or South America, and had not yet joined the continent. Fossil magnetic evidence tells us that Wrangellia was within 15 degrees of the equator before it was added to the continent. By early Jurassic time the Stikine terrane was added onto what is now interior British Columbia. The Alexander and Wrangellia terranes were adjacent to one another by 294 million years ago, in late Paleozoic time, and had joined together off the North American coast by Middle Jurassic time. That was when the rocks of the Gravina Belt were deposited upon this superterrane. Early Cretaceous folding and faulting in Wrangellia terrane rocks mark the initial collision of Wrangellia and North America. Late Cretaceous metamorphism and the intrusion of large volumes of granitic rock into the Coastal Mountain region resulted from this collision. By late Cretaceous time, the Chugach terrane had docked onto North America.

The Alexander terrane includes lower and middle Paleozoic ocean trench sediments and volcanic rocks, shallow water limestone along with limestone, chert, and volcanic rocks formed during late Paleozoic time. During Silurian time, thick sections of shallow water limestone were accumulating in what might have been coral atolls ringing volcanoes that were

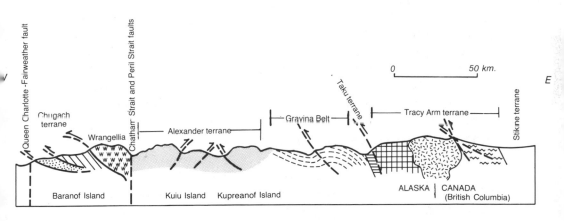

Section across the terranes of southeastern Alaska.

slowly sinking back into the ocean. Thick-shelled clams, reef organisms, and algae were preserved as fossils in limestone that is found today in the Heceta-Tuxekan Islands west of Prince of Wales Island and in Glacier Bay. In places, the limestone rests on boulder beach conglomerate or lavas that were once exposed to the surf.

Triassic volcanic rocks, phyllite, and carbonate and, locally, Jurassic and Cretaceous sandstones and shales overlie the rocks of the Alexander terrane. The terrane is intruded by Jurassic igneous rocks. Broad areas were metamorphosed and thoroughly recrystallized at high pressures and temperatures. To make the picture even more complicated, the rocks have been faulted, segmenting the terrane into smaller blocks. The right-lateral Chatham Strait Fault offset pieces of the terrane about 200 km (120 miles).

The Stikine terrane includes andesite, basalt, and rhyolite lavas and volcanic sedimentary rocks of late Paleozoic age, that are interbedded with marine shales, sandstone, and limestone. On top of these rocks are late Triassic through late Jurassic sedimentary and volcanic rocks, and lower Cretaceous and younger marine and continental volcanic rocks. This terrane probably began as a late Paleozoic volcanic island chain much like Indonesia today.

The Taku terrane is a well-mixed hodgepodge of strata that have been deformed, intruded, and metamorphosed several times. A few fossils of Permian and middle or late Triassic age have been found. Rock types include metamorphosed marine shales and muddy sandstones, metamorphosed andesites, basalts, and rhyolite lavas, along with small amounts of limestone, marble, and conglomerate. The Taku terrane is separated from adjacent terranes by thrust faults to the north and east, and the Chatham Strait Fault to the west. New fossil evidence from the Skagway area suggests that the Taku terrane rocks may be metamorphosed Alexander terrane rocks.

The Tracy Arm terrane contains gneiss, schist, and marble formed through the metamorphism of sedimentary rocks. The metamorphic rocks survive atop the Coast Range batholith, emplaced during several separate igneous events during the last 140 million years. Some geologists now believe this terrane is also a portion of the Alexander terrane, or possibly of the Stikine terrane.

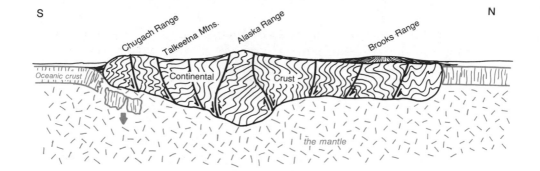

S   N

Section through central Alaska. The crustal blocks are terranes, the faults are the sutures between them.

The Wrangellia terrane is a diverse group of Triassic and late Paleozoic rocks. Triassic basalt, carbonate, and phyllite of the Wrangellia terrane crop out from Cantwell and Chulitna in the Alaska Range, through the Wrangell Mountains, across the Alaska-Yukon border into southeastern Alaska on Chichagoff Island, and even as far south as the Queen Charlotte and Vancouver islands. Some geologists contend that rocks of the Wrangellia terrane exist as far away as the Seven Devils area of west-central Idaho.

The Wrangellia terrane records an action-packed history. This package of rocks was born about 300 million years ago in Pennsylvanian time as an island arc far from the Alaskan coast, perhaps 15 degrees north or south of the equator. As the volcanic eruptions that gave birth to the island arc died out, the arc cooled and sank, subjected to steady erosion by ocean waves. First corals reefs (limestone) accumulated, then deeper water sediments as the terrane continued to sink. The terrane eventually rifted apart about 240 million years ago and the Nikolai greenstone was created by underwater eruptions of basalt. The terrane and its thick load of basalt eventually reached above sea level, but was then eroded down again. Eruptions could not keep pace with erosion. Once again limestone accumulated on the sinking terrane and inevitably, deep water sediments as the terrane submerged beneath deep ocean waters. The Wrangellia terrane, isolated from any continental land mass, continued to drift north until it docked with North America about 120 million years ago.

The Chugach terrane is composed mainly of upper Mesozoic turbidites along with some shale, volcanics, and conglomerates deposited next to an active subduction zone. These slightly metamorphosed sediments are now in contact with a melange of early Mesozoic volcanics, chert, phyllite, and limestone.

Arctic Alaska

During late Jurassic or earliest Cretaceous time an oceanic plate in the Arctic Ocean began to rotate away from the Canadian Arctic Archipelago and then move south. Extension of the oceanic crust in the Arctic Ocean created a rift zone and many small basins. The basins which covered an area several tens of kilometers wide, now underlie the continental shelf between MacKenzie River delta in the Northwest Territories and Point Barrow in Alaska. The basins later filled with marine sediment and are now the main targets for petroleum exploration in the north. The rotating Arctic Ocean plate pushed against northern Alaska, giving birth to the Brooks Range.

With uplift, erosion outpaced deposition and older sediments were partly eroded. In both the subsurface rocks of the Prudhoe Bay area and in surface outcrops of the Arctic National Wildlife Refuge, a major unconformity near the base of Cretaceous rocks truncates all preexisting rock units. The unconformity controls the location of oil in Prudhoe Bay; it is overlain by rich organic shales that seem to be the source rocks for the hydrocarbons. By late Cretaceous time, dinosaurs were using this coastal habitat (see page 223).

The Brooks Range formed along the compressed margin of a plate formed as the Arctic Basin was opening up by plate rotation. The suture making the plate collision lies at the surface along the Kobuk River. Underthrusting the oceanic plate peeled slices of Paleozoic and early Mesozoic sedimentary rocks off the subducted slab, stacking them against the continent. Ore deposits in the southern Brooks Range are probably related to volcanoes that existed before the formation of the modern range.

Alaska in Tertiary Time

About 65 million years ago, the Chugach terrane was colliding with Alaska and North America. A mountain range north of the collision zone probably included active volcanoes. Granite that formed during the episode is exposed now at the surface in the modern Talkeetna Mountains north of Anchorage. The collision of the Chugach terrane with Alaska may have resembled the collision now taking place between the Yakutat block and the St. Elias Mountains. Continental thickening accompanying this modern collision causes continued uplift of the St. Elias Mountains, along with rapid rates of erosion and delivery of sediments to adjacent valleys and the continental shelf. The collision has generated magma for volcanoes in the Wrangell Mountains and for granitic plutons miles below the surface of the volcanoes.

In Eocene time the Prince William terrane arrived and docked against southern Alaska. About the same time, 42 million years ago, there was a major change in the motion of the Pacific Plate. The shift is documented by a trail of well-known ocean floor volcanoes which includes the Emperor Seamount chain and the Hawaiian Islands. The bend between the northern Hawaiian Islands and the southwestern Emperor Seamount marks the change in plate direction. Terranes arriving in Alaska before 42 million years ago collided head-on. Those arriving after that were added by side-slipping along the coast. Presently the Pacific plate slides northwestward past the west coast of North America and is consumed by subduction along the Aleutian trench.

Formation of the Bering Sea Shelf

The Bering Sea shelf is a large shallow-water shelf that borders the area north of the Aleutian Islands between western Alaska and the eastern-most arctic regions in the Soviet Union. It contains the only surviving piece of the Kula plate, a piece that was rescued from destruction when the Kula plate jammed along its subduction zone and a portion of it became fused to Alaska. The name Kula is an Athabaskan word meaning "all gone" and the Kula plate would indeed be all gone if it weren't for the remnant in the Bering Sea.

During much of Mesozoic time the Eurasian-North American plate was overriding the Kula plate. By middle Eocene time, between 42 to 44 million years ago, a change in the direction of motion of the Pacific plate from northward to northwestward jammed the Kula plate against North

America. A new subduction zone formed along the former spreading center that separated the Kula and Pacific plates. The shift effectively stranded the unsubducted piece of the Kula plate in the Bering Sea, forming a piece of the Bering Sea shelf.

Extensional basins formed later in Tertiary time on the Bering Sea shelf. The Navarin and St. George basins formed during this time and were then filled with Tertiary marine sediments. These two basins are like oil-rich salty basins in the Gulf of Mexico that formed under similar circumstances on a large stranded marine plate fragment. The Bering Sea shelf will be an important area for oil exploration.

Cook Inlet Tertiary Sediments

During Tertiary time terrestrial sediments accumulated throughout Alaska. In Cook Inlet, during Paleocene time, at least 5,000 feet (1500 meters) of sediments accumulated to form the Chickaloon Formation. Thick beds of conglomerate record a time of rapid erosion of mountains. Miocene plant fossils in Cook Inlet represent a diversified warm-temperature flora, from a time of warmer world climate when sediments accumulated in broad deltas along a more subdued Alaskan coast. Younger sediments tell the story of a gradually cooling climate, culminating in the geologically recent Ice Ages.

Pleistocene Glaciation in Alaska

The St. Elias Mountains and other ranges adjacent to the Gulf of Alaska probably supported mountain glaciers as early as 10 to 13 million years ago. Frequent north Pacific storms furnish such a plentiful supply of moisture that the southern coast of Alaska has remained under glacial ice for at least the last 5 million years. Late Miocene glacial mudstone in the Yakataga Formation along the Alaskan Gulf was produced by a "rain" of silt, sand, and gravel pouring onto the continental shelf from floating icebergs calved off tidewater glaciers.

Mountain ice first appeared in the Brooks and Alaska ranges during late Tertiary time. Four major glacial periods are recognized from glacial deposits in the central Brooks Range. At least six episodes of glaciation have been identified in the western Alaska Range. Some of these glacial deposits have been offset by active branches of the Farewell fault.

Interior Alaska north of the Alaska Range and south of the Brooks Range was never glaciated. The exception is the Yukon-Tanana Upland west of Fairbanks where a record of six glacial episodes, one of them possibly late Tertiary, is recognized. The cold climate restricted lowland forests, while the broad floodplains in front of the glaciers supported grassland vegetation suitable for a great number of grazing animals such as bison, horse, and mammoth. When this vegetation changed at the end of the Pleistocene these mammals disappeared. North of the Brooks Range an ice-free polar desert existed. Large dune fields developed as glacial sand was transported by the wind. The middle Kobuk Valley near the village of Ambler, presently contains a large area of active dunes covering an area of about 130 km. The Kobuk River continues to drain areas that were glaciated during the last Ice Age.

During periods of maximum glaciation, sea level was lowered and shallow portions of the continental shelves were dry land. The Bering Sea shelf acted as a land bridge between Alaska and the USSR during these intervals, allowing a wide range of animals and early Alaskans to wander freely back and forth. By 40,000 years ago human hunters shared northern habitats with mammoths from France to Siberia. Many scientists now believe these people followed the mammoths across

During the last ice age, sea level was about 300 feet lower than it now is. Large areas that are now under water were then high and dry.

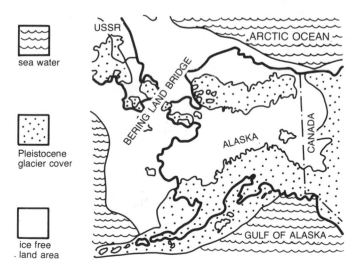

sea water

Pleistocene glacier cover

ice free land area

the land bridge, leaving the Old World behind.

All along the southern coast of Alaska, the mountains were shaped by intermontane ice sheets and alpine glaciers a number of times during the Pleistocene. Unfortunately any evidence of late Pleistocene glacial erosion and deposition was later disguised or destroyed by younger glacial expansions. Many of southeast Alaska's glacial features may be the result of very recent glaciations.

A warming of the climate caused a general retreat of late Pleistocene ice that ended in a "thermal maximum" 6 to 7 thousand years ago. At that time Alaska glaciers were reduced to their present size or a little smaller. Rising sea level, caused by the melting of great volumes of ice at the poles and from the continental ice sheets, flooded many of the glacial valleys and carved marine terraces at several levels above present tideline. In southeast Alaska, marine terraces and beach deposits occur at elevations of 450 feet on the mainland west of Prince of Wales Island and at about 180 feet near both Petersburg and Ketchikan. On Admiralty Island outcrops of glacial-marine sediment at 639 feet record the flooding of the island by rising sea level.

Between 2,000 and 3,000 years ago the climate began to cool and many glaciers advanced. In Glacier Bay, an ice sheet grew to a thickness of 4,000 feet and advanced into Icy Strait. The ice sheet fluctuated during that time and began a retreat about 230 years ago when early European explorers began to visit the region and keep records. Studies of tree rings growing on or near glacial moraines in the Juneau area have shown that local glaciers reached their maximum positions by around 1750. This advance is marked by prominent terminal moraines in front of Herbert, Eagle, and Mendenhall glaciers. Today most glaciers are in retreat and sea level is slowly rising. There are exceptions however, and sometimes minor readvances occur such as at the Taku, Brady, Lituya, Hubbard, and Harvard glaciers.

Regional Uplift Following Deglaciation

In some coastal areas the emergence or elevation of shoreline is recorded by newly exposed shoals, raised beaches and mud flats, and wave-cut benches. Between 1959 and 1960 uplift of the land between Yakutat and Petersburg was measured. The greatest uplift rates measured were 1.6 inches/year. This is a recently deglaciated area and the removal of many tons of ice takes great weight off the earth's crust, which rides upon the earth's interior much like an ice berg floating on water. When glacial ice lay in thick sheets on the land the additional weight caused rock to deform plastically beneath the surface and to flow away from ice accumulation centers where the earth's surface had been pushed down much like the sag in a mattress under the weight of a heavy sleeper. When the ice later melted, it was removed more rapidly than the subcrustal rock could flow back. The uplift that is currently going on is due to subcrustal, plastically-behaving rocks flowing back in response to the removal of ice at the surface. The rate of rebound and the size of the area involved help geologists to estimate the thickness of ice and to determine how great an area was covered by ice.

Map Symbols for the Southeast Panhandle

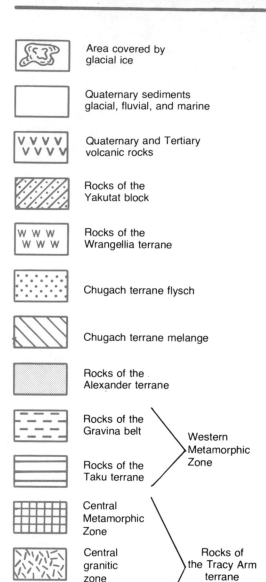

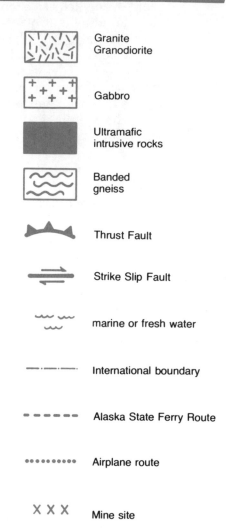

Area covered by glacial ice

Quaternary sediments glacial, fluvial, and marine

Quaternary and Tertiary volcanic rocks

Rocks of the Yakutat block

Rocks of the Wrangellia terrane

Chugach terrane flysch

Chugach terrane melange

Rocks of the Alexander terrane

Rocks of the Gravina belt

Rocks of the Taku terrane

Western Metamorphic Zone

Central Metamorphic Zone

Central granitic zone

Eastern Metamorphic Zone

Rocks of the Tracy Arm terrane

Granite Granodiorite

Gabbro

Ultramafic intrusive rocks

Banded gneiss

Thrust Fault

Strike Slip Fault

marine or fresh water

International boundary

Alaska State Ferry Route

Airplane route

X X X   Mine site

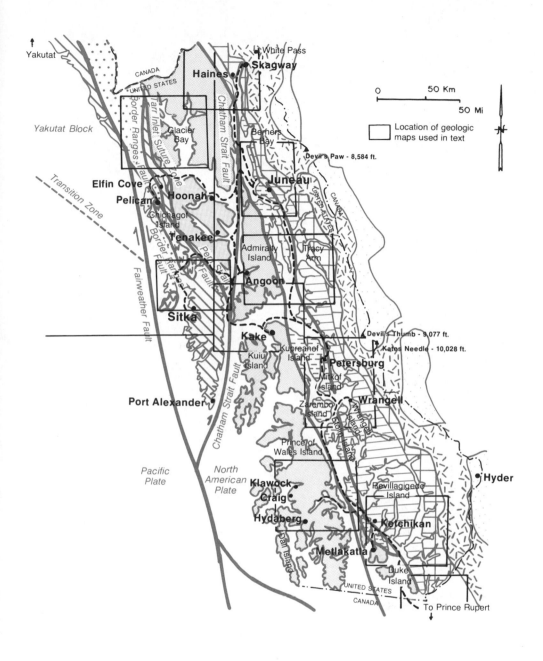

Index to geologic maps of southeast Alaska.

19

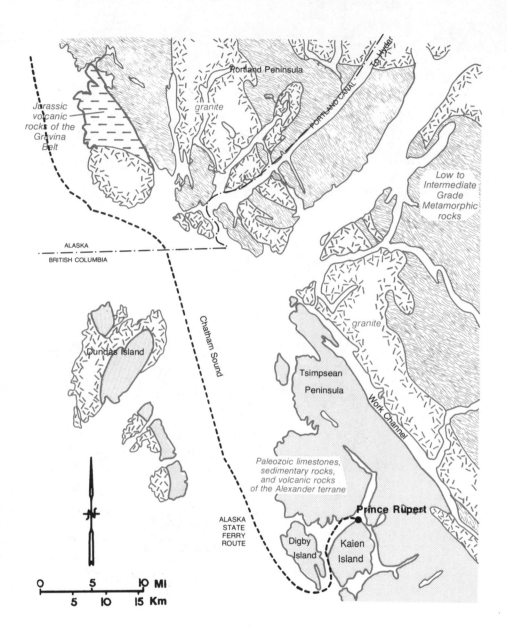

Jurassic volcanic rocks of the Gravina Belt

granite

Portland Peninsula

To Hyder

PORTLAND CANAL

Low to Intermediate Grade Metamorphic rocks

ALASKA
BRITISH COLUMBIA

Dundas Island

Chatham Sound

granite

Tsimpsean Peninsula

Work Channel

Paleozoic limestones, sedimentary rocks, and volcanic rocks of the Alexander terrane

Prince Rupert

ALASKA STATE FERRY ROUTE

Digby Island

Kaien Island

0   5   10 MI
5   10   15 Km

PRINCE RUPERT REGION
Ferry Route

I
SOUTHEAST ALASKA

INTRODUCTION

Southeast Alaska is a geologically complex region about half the size of the state of California. With its many offshore islands it is known as the Alexander Archipelago. This region measures 125 by 400 miles and lies at the same latitudes as Scotland, Denmark, and northern Sweden. The thousands of forest-covered islands are in essence a northern rainforest. The terrain is mountainous and no roads connect the major cities. The geology of this area is described as one might see it sitting in the observation lounge of one of Alaska's comfortable ferries, travelling the marine highway from Prince Rupert, British Columbia to Skagway, Alaska.

THE ALASKA MARINE HIGHWAY

Prince Rupert—Ketchikan
Sailing time: 6 hours 15 minutes
90 miles

From Prince Rupert, British Columbia the Alaska State ferry sails southward between Digby and Kaien islands and then turns northward upon rounding Digby Island to enter Chatham Sound. Our story begins as the ferry reaches the International Boundary about 54 degrees 40' Latitude. Portland Canal to the east forms part of the boundary between Alaska and British Columbia. Named by George Vancouver who in 1793 was in the area searching for the fabled northwest passage, the canal is actually a drowned glacially carved valley or fjord. It extends 100 miles northeast to the old mining town of Hyder, a silver-rich area where miners continue to work underground lode deposits. Mines near Hyder shipped or stockpiled a few hundred tons of ore or concentrates of tungsten copper, lead, and zinc minerals.

Portland Peninsula

The Portland Peninsula to the east is bounded by Behm Canal, Revillagigedo Channel and Portland canal. Except for small outcrop areas of Quaternary or Tertiary basalt and andesite, two types of lavas, the bulk of this peninsula is underlain by a group of Mesozoic and Cenozoic plutons or igneous intrusive rocks that are called the "Coast Plutonic-Metamorphic Complex."

Those sedimentary rocks were metamorphosed under conditions of high temperature and pressure to form metamorphic rocks called amphibolite and granulite. The original ages and environments of deposition of the sedimentary rocks are obscured by their present metamorphic form. The rocks occur within or between the plutons as dividing walls, roof remnants, and inclusions. The igneous rocks cooled within magma chambers at depths of roughly 4 to 8 miles beneath the earth's surface. Uplift and erosion have elevated them to the surface for ferryside geologists.

Revillagigedo Channel

As the ferry enters Revillagigedo Channel, Duke Island is to the west. This island is noted for its Upper Triassic sequence of rhyolite, basalt, and sedimentary rocks, along with a Cretaceous zoned ultramafic intrusive complex that shows spectacular sections of rhythmically layered dunite and peridotite, rocks rich in iron, magnesium, and titanium minerals. Dunite consists almost entirely of the green mineral olivine. Peridotite contains olivine and pyroxene. Other minerals present here are barite, and ores of copper, lead, and zinc. Gold and silver exist in volcanic sulfide deposits in the upper Triassic rocks, and titanium-magnetite deposits are in the ultramafic rocks.

Quartz Hill Molybdenum Deposit

Farther north up Revillagigedo Channel, the ferry passes Boca de Quadra Fjord and then Behm Canal. These two waterways lead to the boundary of Quartz Hill, one of the world's largest molybdenum deposits. The deposit is 65 miles east of Ketchikan, in Wilson Arm of Smeaton Bay, an inlet off Behm Canal. Molybdenum is an important mineral used for the manufacture of high-strength steel, for high-temperature alloys, and for corrosion-resistant materials. The Quartz Hill deposit is in stocks and dikes formed by small masses of magma that were cooling at depths less than four miles beneath the earth's surface. Molten rock intruded into the older plutonic rocks of the Coast Range Mountains after the region had experienced uplift and erosion. The molybdenum-bearing rocks have been dated at between 27 and 30 million years old (Oligocene). The ore is within a small igneous intrusion known as the Quartz Hill stock. The molybdeunum mineral forms metallic, bluish gray coatings along fractures in the rock and also occurs in veins with quartz.

There are plans to develop the deposit into a mine to produce from a zone that contains about 1.5 billion tons of low-grade molybdenum ore. To mine the ore the company would like to dig an open pit over two miles long and 1800 feet deep. Disposal of the mine tailings will be a big job. Fishermen and conservationists are concerned that the mine sediments will harm the local shrimp, herring, crab, and salmon fisheries and would like the tailings dumped into the middle of the deep Boca de Quadra fjord. The mining company, however, is concerned about the cost of mining the ore and prefers to dump the tailings in inner Boca de Quadra or Smeaton Bay.

Fossil coral from Alexander terrane rocks on Hotspur Island northwest of Duke Island.

—U.S. Geological Survey photo by T. Chapin

Pillow lava from Alexander terrane rocks on the west coast of Gravina Island. —U.S. Geological Survey photo by T. Chapin

Gravina Island

The region surrounding Revillagigedo Channel is composed of several distinct terranes that docked against the North American plate and fused together. On Gravina Island across the channel from Ketchikan, high angle north-northwest-trending faults divide the island into two metamorphic provinces. The rocks on the western half of the island are part of the Alexander terrane. They are only slightly metamorphosed.

Rocks on eastern Gravina Island, including those underlying the airport, have been metamorphosed to greenschist by exposure to high pressures and low temperatures near cooling batholiths deep within the crust. These rocks are part of the Gravina belt and were deposited upon the Alexander-Wrangellia terrane before it joined North America. They include lower Cretaceous to mid-Jurassic muddy sandstones and dark volcanic rocks that contain intrusions of dark igneous rocks. In southwestern Gravina Island the Gravina belt rocks are faulted against the Alexander terrane rocks.

Gold and silver mines were once active on Gravina Island, at Helm Bay on the Cleveland Peninsula, along Tongass Narrows and at Thorne Arm on Revillagigedo Island. The mines on Gravina Island also shipped small amounts of copper ore.

Revillagigedo Island and Ketchikan

A fault along Tongass Narrows separates Gravina and Revillagigedo islands. It split a small gabbro pluton of Miocene or Oligocene age, and moved the pieces about four miles.

Rocks east of Tongass Narrows on Revillagigedo Island and the Cleveland Peninsula to the north are a varied group. Magmas intruded metamorphosed and tightly folded sedimentary and volcanic rocks of Permian and middle Triassic age. These rocks are considered to be part of the Taku terrane, but may have been deposited at the same time as similar rocks in the Gravina belt. The largest mass of igneous rocks is mid-Cretaceous in age, and extends from northern Revillagigedo Island to the adjacent Cleveland Peninsula. Bell Island Hot Springs, a favorite recreating spot for locals and visitors alike, gets its heated waters from rainwater that descends deep beneath the surface via fractures and faults to be heated by deep igneous rocks. The heated water then rises buoyantly to the surface.

The sedimentary and volcanic rocks were folded, the folds pushed over to the southwest, severed by thrust faults. Stresses along the fractured surfaces created a type of metamorphic rock known as mylonite, a banded or streaky rock that forms through shearing of rocks that have been smeared along fault zones.

On southwestern Revillagigedo Island older Permian marble has been shoved over younger Triassic bedded rocks along a thrust fault. The northeastern third of Revillagigedo Island grades from slightly metamorphosed rocks, greenschists, to thoroughly metamorphosed

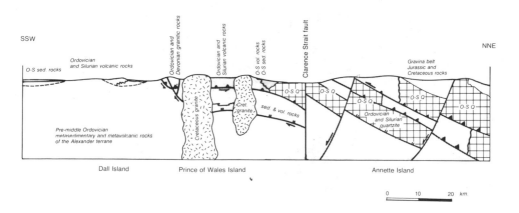

Section through metamorphic rocks of the Alexander terrane. Numerous faults have cut and shuffled blocks within the terrane. Granitic intrusion followed docking of the terrane.

25

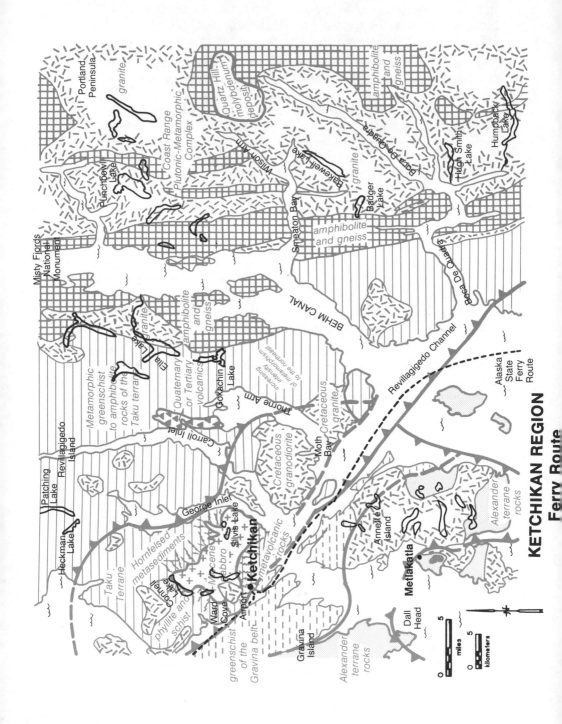

KETCHIKAN REGION
Ferry Route

Portland Peninsula

granite

Coast Range Plutonic-Metamorphic Complex

Quartz Hill molybdenum deposit

amphibolite and gneiss

Boca De Quadra

Hugh Smith Lake

Humpback Lake

Punchbowl Lake

Misty Fjords National Monument

Wilson Arm

Bakewell Lake

Sneaton Bay

granite

Badger Lake

amphibolite and gneiss

Boca De Quadra

BEHM CANAL

Revillagigedo Channel

Alaska State Ferry Route

Revillagigedo Island

Metamorphic greenschist to amphibolite rocks of the Taku terrane

Ella Lake

granite

Quaternary or Tertiary volcanics

amphibolite and gneiss

Gokachin Lake

increasing metamorphism to the northeast

Thorne Arm

Moth Cretaceous Bay granite

Cretaceous granodiorite

Alexander terrane rocks

Annette Island

Patching Lake

Carroll Inlet

George Inlet

Heckman Lake

Taku Terrane

Hornfelsed metasediments

Silvis Lake

Ketchikan

metavolcanic rocks

Metlakatla

Alexander terrane rocks

Dall Head

Connell Lake

Ward Cove

Miocene gabbro

Airport

phyllite and schist

greenschist of the Gravina belt

Gravina Island

Alexander terrane rocks

5 miles

5 kilometers

schists and gneisses. The remainder of the island consists of green-schist rocks recrystallized at temperatures between 300 and 500 degrees Centigrade. This metamorphism happened during the middle or late Cretaceous time, and probably records the arrival of terranes that joined North America. The collision of terranes created the pressures and temperatures necessary to cook, fold, and fault the rocks into their present jumbled state.

Ketchikan Roadside Geology

Between the ferry terminal and Ketchikan, the highway passes through a tunnel of metamorphosed volcanic and sedimentary rocks that are also exposed on the lower ridge east of the downtown area. These rocks are late Paleozoic to early Mesozoic in age. They include dark gray, silvery-green, and greenish-gray phyllite and schist along with small amounts of marble.

Some outcrops of the volcanic rocks look identical to Gravina belt rocks of Jurassic, Cretaceous, and late Triassic age on Annette and Gravina islands. Look for the metamorphosed rocks in the cliff east of Tatsuda Market, just south of Creek Street.

Deer Mountain at an elevation of 3000 feet dominates Ketchikan. A hike to its 1500 foot elevation provides a viewpoint of Tongass Narrows and Annette Island. Hiking through the metamorphosed volcanic rocks toward the peak, roadside geologists can see to the north and west across the Tongass Narrows fault separating the Taku terrane of Revillagigedo Island from the Gravina belt rocks and the Alexander terrane of Gravina Island, Annette Island and Prince of Wales Island to the west. Two miles past the summit along the trail to Blue Lake, is Granite Basin, where the rocks are actually gabbro, the intrusive form of basalt.

Late Paleozoic to early Mesozoic schist and phyllite form the rock walls of this tunnel in downtown Ketchikan. —CC

Schist at Rotary Beach along the South Tongass Highway. —CC

South Tongass Highway

Along the South Tongass Highway, Mile 3.5 at Rotary beach provides large boulders of metavolanic rocks sitting on schist. Crystals of
amphibole and well-formed cubes of pyrite or "fool's gold" occur in
these boulders.

Farther south on the South Tongass Highway, the Tertiary gabbro
pluton appears in the road cuts. Gabbros are coarse-grained, dark
igneous rocks that have the same chemical components as basalt but
cool slowly in deeper magma chambers to form visible crystals.
Basalts chill quickly at the earth's surface after a volanic eruption
and are not made up of large, visible crystals. Reddish piles of sand
developed from the weathering of the fractured gabbro can be seen at
the base of the cliffs. Groundwater moving through the rock along
these cracks is able to oxidize or rust the iron-bearing minerals in the
gabbro, especially the mineral magnetite. Surviving core boulders of
unweathered gabbro are left surrounded by sandy weathering pro-
ducts. At mile 10.5 the younger gabbro can be seen in contact with
and intruding the Paleozoic-Mesozoic metavolcanic rocks. These relationships in outcrops enable geologists to figure out which is the
older of two rock units.

From the end of the South Tongass Highway there is hiking access
to Twin Peaks. Start up from Beaver Falls Power Station. From the
North Peak at an elevation of 2,880 feet or the South Peak at an
elevation of 3,090 feet, there are spectacular views southeast across
Carroll Inlet into the Misty Fjords National Monument and the rocks
of the Coast Range Metamorphic-Plutonic Complex.

Upper Paleozoic to Mesozoic phyllite is exposed near the end of the pavement on the North Tongass Highway. —CC

North Tongass Highway

About two and one-half miles north of Ketchikan, red soil and phyllite or schist are exposed in a quarry just north of the post office and ferry terminal. Here large quartz boulders from veins in the country rock decorate a nearby garden.

The metavolcanic rocks exposed in downtown Ketchikan grade into schistose metamorphic rocks that are exposed along the North Tongass Highway and crop out especially well at Settler's Cove at mile 18.2 at the north end of this road. These rocks were late Paleozoic and early Mesozoic muds, volcanically-derived sediments, and andesites and basalts that were later folded and intruded by quartz veins. They contain quite an assortment of minerals: quartz, feldspar, biotite, garnet, muscovite, pyrite, graphite, and hornblende. Locally the schist carries abundant pyrite and minor chalcopyrite, and has been prospected for copper, gold, and other metals. Frequent outcrops of the schist occur along the North Tongass Highway and the schist has been quarried near the end of the paved road to make building sites for houses in a nearby subdivision.

At the turnout opposite Guard Island, Cretaceous granodiorite with large crystals or phenocrysts is exposed in the roadcut on the uphill side. The large crystals are plagioclase feldspar, some over an inch long. Garnet occurs here as well-formed, small crystals in the cores of the plagioclase crystals. This intrusive rock has been dated at 80 to 87 million years old.

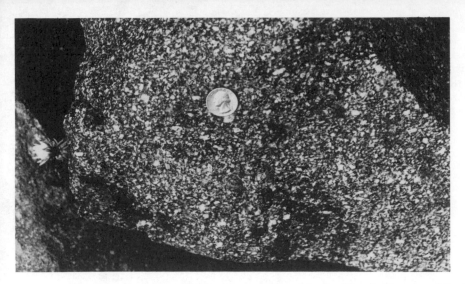

Cretaceous granodiorite outcrops beside the North Tongass Highway opposite Guard Island turnout. —CC

Ward Lake Road

The road leaves the coastline at Ward Cove and winds through metamorhpic schists that were intruded by the Miocene or Oligocene gabbro complex to the east. In places the bedrock ridges are glacially polished and grooved. Along the private logging road past the junction with Harriet Hunt Road, clear-cutting has exposed bedrock that contains amphibole and mica near the northern edge of a gabbro pluton.

Ketchikan—Wrangell
Sailing time about 6 hours
85 air miles

As the ferry departs from Ketchikan to head north up Tongass Narrows, passengers can see greenish rocks exposed on Gravina Island below the airport runway. This rock, which extends along the shoreline to the northern tip of Gravina Island, is greenschist—metamorphosed andesite and basaltic ash and agglomerate. Small amounts of grayish-green and black phyllite and phyllitic siltstone intertongue with the metavolcanic rocks. The age of metamorphism is Cretaceous, if we can rely on dates from similar rocks elsewhere in the Ketchikan and Prince Rupert region.

Approximately one-half hour after departing from Ketchikan the ferry passes the northern tip of Gravina Island. Tongass Narrows, the waterway separating Gravina and Revillagigedo islands, follows a fault that horizontally offset the Miocene or Oligocene gabbro pluton east of Ketchikan about 4 miles. This fault also separates rocks of the Gravina belt on the west from rocks of the Taku terrane on the east.

The ferry next enters Clarence Strait, a valley formed initially by the Clarence Strait Fault, later enlarged by the carving action of glacial ice, and finally drowned by rising sea level at the end of the last Ice Age. To the west is the northern Kasaan Peninsula on Prince of Wales Island and to the east is the Cleveland Peninsula, an extension of the mainland.

Glacial Deposits Near Hollis

Two deposits of glacial till from eastern Prince of Wales Island record two glacial advances. The older till was left by a glaciation that extended to elevations above 3,000 feet. The younger till was left by a

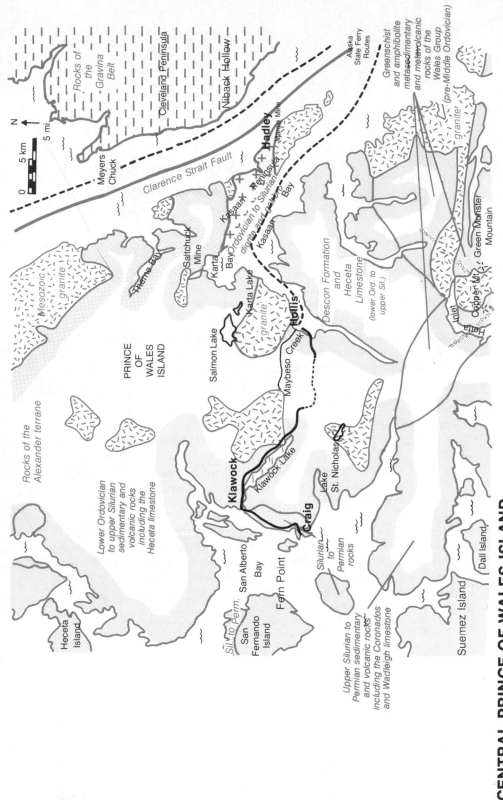

CENTRAL PRINCE OF WALES ISLAND
Ferry Route

Rocks of
the
Gravina Belt

Cleveland Peninsula

Niback Hollow

Alaska State Ferry Routes

Greenschist and amphibolite metasedimentary and metavolcanic rocks of the Wales Group (pre-Middle Ordovician)

N

5 mi

5 km

Meyers Chuck

Hadley

Marble Mine

Clarence Strait Fault

Kasaan Peninsula

Kasaan

Ordovician to Silurian diorite and gabbro

Kasaan Bay

granite

Mesozoic granite

Theme Bay

Saltchuck Mine

Karta Bay

Descon Formation and Heceta Limestone
(lower Ord. to upper Sil.)

Copper Mt.

Green Monster Mountain

Hetta Inlet

PRINCE
OF
WALES
ISLAND

Karta Lake

Karta Lake granite

Hollis

Salmon Lake

Maybeso Creek

Rocks of the Alexander terrane

Klawock

Klawock Lake

Lake St. Nicholas

Lower Ordovician to upper Silurian sedimentary and volcanic rocks including the Heceta limestone

Craig

San Alberto Bay

Fern Point

Silurian to Permian rocks

Heceta Island

San Fernando Island

Sil. to Perm.

Upper Silurian to Permian sedimentary and volcanic rocks including the Coronados and Wadleigh limestone

Suemez Island

Dall Island

A glaciated Silurian mudstone surface near Karta Bay, Prince Wales Island provided a good carving site for an early Alaskan native.

—U.S. Geological Survey
photo by T. Chapin

glacier that apparently moved down the Maybeso Creek valley near Hollis from the central mountains on the island. It is found at altitudes lower than 1500 feet and forms a group of moraines in the valley. Sediments overlying and younger than the youngest till have been dated about 9,510 years, so the younger glaciation must have happened before then.

Metal Deposits and Epidote on Prince of Wales Island

The Salt Chuck Mine at the head of Kassan Bay is a copper-palladium mine. The copper and platinum minerals (palladium is one of the platinum group of metals) are found along the contact between pyroxenite and gabbro. Platinum is an important metal used to make catalytic converters for pollution control devices on cars. For many

1915 photo of the Mamie Mine on the Kasaan Peninsula, Prince of Wales Island. Iron was mined from a skarn deposit formed by the intrusion of granitic rocks into limestone.

—U.S. Geological Survey
photo by T. Chapin

33

years copper was second only to gold in value of production in south-eastern Alaska. Near Hetta Inlet, the copper ore formed in skarns where granitic plutons intruded the limestone of the Alexander terrane.

At Green Monster Mountain at the northeast end of Hetta Inlet in south Prince of Wales Island, the skarn zone next to Copper Mountain yielded a world-class specimen of epidote. This speciman is about 8 inches long and was purchased by the University of Alaska where it now resides in the mineral collection of the Fairbanks campus. Epidotes from Green Monster Mountain are found in museum collections all over the world. The beautiful, large, dark greenish-black crystals find rivals only in epidotes from a now exhausted mine at Unter-sulzbachtal, Austria. Other collector-quality epidotes that come from a mine in Russia are not as good as these from Alaska. These Green Monster and Copper Mountain epidote claims are privately owned and closed to collectors.

Marble and Limestone

Marble and limestone on western Prince of Wales Island and neighboring islands are part of the Alexander terrane. The marble and limestone are most abundant in rocks of middle Silurian and younger Paleozoic sections. The Silurian limestones in the Heceta-Tuxekan Islands area just off the western coast of Prince of Wales Island are more than 10,000 feet thick. They record a long history of deposition within intertidal and beach environments.

Around 1900 there was a great demand for ornamental and building stone along the west coast of the continental U.S. Shipments of high quality ornamental marble began in 1902 from quarries at Tokeen on Marble Island, the largest of the marble quarries that operated in southeastern Alaska between 1909 and 1932. At least 2,150,000 tons of chemical-grade stone and 450,000 tons of structral grade stone have been quarried from a dozen sites on Prince of Wales and Dall islands. The principal marble quarries, however, were in the Tokeen-Calder area on Kosciusko and Marble islands.

The Tokeen marble quarry circa 1916 on Marble Island northwest of Prince of Wales Island.
—U.S. Geological Survey photo by A.H. Brooks

34

The state capitol in Juneau has four large exterior columns of Tokeen marble, as well as marble wainscoting and trim inside. Inflation and change in building styles put an end to the southeast Alaska marble industry after the Second World War. An estimated 800 million tons of high quality marble remain in the southeast Alaskan quarries awaiting the day when marble comes back into vogue.

Groundwater draining through acid bogs in alpine muskegs on Northern Prince of Wales Island has seeped into and widened vertical fractures in the limestone and developed an extensive system of caves. In the summer of 1991, the bones of an exceptionally large grizzly bear (*Ursus arctos*) was found in El Capitan Cave. Recent dating of the bones has yielded an age of about 9,760 years. This information suggests that grizzly bears occupied the island after glacial ice had melted back to the mainland, and subsequently the bears vanished as they are no longer found on Prince of Wales Island. These are the first Pleistocene vertebrate fossils recovered from southeast Alaska.

Etolin Island

Two hours and twenty minutes north from Ketchikan, the ferry passes Onslow Island to the east, completing its crossing of Ernest Sound. Etolin Island is now east of the ferry and Prince of Wales Island is on the west.

The Screen Islands, just south of Steamer Knoll on Etolin Island are the easternmost rocks of the Alexander terrane. The Screen Islands rocks match the Upper Triassic Burnt Island Conglomerate of Keku Strait northwest of Kupreanof Island. They include conglomerate, calcareous sandstone, calcareous siltstone, and limestone. Many fossils have been collected from Screen Islands. Some of the cobbles of white chert in the conglomerate contain late Permian brachiopods, or lamp shells. Interlayered sandstone and limestone overlying the conglomerate contain ammonites, a now extinct relative of today's chambered nautilus, and early Mesozoic brachiopods. Cousins of this type of brachipod are common today, attached to bedrock or boulders living in the subtidal waters of Alaska and the Pacific Northwest. The Screen Islands sediments also contain a late Triassic swimming scallop and conodonts, tiny tooth or jaw-like fossils. Fossil-bearing chert cobbles of Permian age in the Triassic conglomerate were probably reworked from an older source.

It is possible that these sediments accumulated in a small basin adjacent to active volcanic islands. Sedimentary structures in the conglomerate and sandstone show that the rocks were deposited near shoreline and that the environment became quieter and deeper by the time of limestone deposition. Fossils in the limestone represent deep-water swimming animals. The pale color of the fossil conodonts indicates burial temperatures of 110 to 200 degrees C, unusually low for this part of Alaska.

Fossil magnetic data from rocks on Hound Island in Keku Strait show that the Alexander terrane has not moved northward with respect to North America since late Triassic time.

Three hours and twenty minutes north of Ketchikan the ferry passes Steamer Knoll, on northwestern Etolin Island and its neighbor, tiny Marsh Island.

Northwestern Etolin and Marsh Islands

Glaciated 20 million year old knobs of granite lie south of Quiet Harbor on northwestern Etolin Island. —CC

Upper Mesozoic rocks of the Gravina belt exist on Etolin Island on the east or starboard side of the ferry. The rocks on the western side of the island are folded and faulted, but so slightly metamorphosed that original sedimentary beds and volcanic textures are preserved. The youngest layered rocks are on Marsh Island, just offshore of western Etolin near Steamer Knoll. A 50 foot outcrop of thinly interbedded mudstone, stiltstone, and sandstone made up of volcanic material, and andesite ash contains fossil ammonites, approximately 100 million years old. Beneath these Cretaceous beds are stiltstone, mudstone, and a bit of conglomerate and limestone. These late Jurassic to early Cretaceous beds have fossils of *Buchia*, a small mussel-like mollusk that lived in shallow sea water during Mesozoic time.

On top of the lower Cretaceous sediments on Marsh Island is a thick pile of volanic rock that includes tuff, flows, and volcanic sandstones, as well as lenses of mudstone. Evidently volcanoes erupted in this region during early Cretaceous time.

Central and eastern Etolin Island, Woronkofski, Wrangell, eastern Zarembo, Mitkof and Kupreanof Islands all consist mainly of granitic plutons. They also contain small outcrops of fossil-free phyllite, greenschist, and metamorphosed andesite. These rocks are probably the metamorphosed relatives of the rocks exposed on Marsh Island. The light-colored granites within the Kuiu-Etolin volcanic-plutonic belt are 19 to 23 million years old. North of here the ferry enters Stikine Strait, turning eastward toward the town of Wrangell with Zarembo Island to the northwest. Massive granite along the southeastern Zarembo Island coastline weathers into blocks along fractures.

As the port of Wrangell comes into view, the enormous Stikine River delta can be seen seven miles to the northeast. The huge expanse of sand flats and winding channels represents the endpoint of the great volume of sediment the river carries into Dry Strait. To avoid running aground, the ferry will not pass by the Stikine delta to travel northward to Petersburg, but will backtrack to Wrangell Narrows. The Stikine River delta is one of the most productive estuaries in southeast Alaska. It is a rich nursery for fish and shellfish, a place where huge concentrations of smelt run annually, and a stopover for more than half a million birds migrating through each fall and spring.

Stikine River Basin

The ancestral Stikine River may have flowed from interior British Columbia out to the southeastern Alaska coast beginning about 50 million years ago, before uplift of the Coastal Mountains. The present 400 mile long channel was established during the past 7 million years as the river eroded through the uplifting Stikine Plateau and Coast Mountains. The river drains an area of 20,000 square miles; it has its headwaters in the Cassiar Mining District of northern British Columbia. This naturally formed waterway through the coastal mountains has been an important thoroughfare for migrating wildlife and early Alaskans. The Tlingit people travelled down river from the

Rippled sands mark the Stikine River delta at low tide. View from southern Mitkof Island. —Rod Flynn photo

interior and made their living as a seafaring people. For many centuries they returned upriver each summer to catch salmon, pick berries, and trade with the Tahltans, an Interior Athabaskan group.

The explorer George Vancouver passed by while preparing a chart of the coastline from California to Alaska in 1793 but did not recognize the river. Beginning in 1861, and continuing through 1873-74, waves of prospectors seeking gold in the Cassiar District moved upriver from Wrangell during the Cassiar Gold Rush. When word of gold discoveries in the Klondike reached the outside world between 1897-1898, the Canadians developed a scheme for a railway up the Stikine to the Klondike via interior British Columbia. This plan was killed in Parliament and the traffic diverted north to Juneau and the Chilkoot Pass.

Gold was the main attraction of the Stikine region in the 1800s. Today, copper and other minerals draw mining interests to the area. Some companies estimate that 130 million tons of copper-gold-silver ore exist at Galore Creek, between Telegraph Creek and Wrangell. From Dease Lake, British Columbia, the Kutcho Mining road leads to undeveloped deposits of metal sulfides, gold, jade, and asbestos.

On the interior side of the Coast Mountains, east and south of Telegraph Creek is British Columbia's equivalent of Katmai and the Valley of Ten Thousand Smokes, Mount Edziza Provincial Park. Mt. Edziza (9,143 feet) is surrounded by the pockmarked Klastine Plateau, a volcanic landscape marked recently by Cocoa and Coffee Craters. The volcanic activity changed the courses of several surrounding rivers and blanketed adjacent mountains with ashfall.

The Edziza Volcanoes lie northeast of the Coast Range and are Miocene to Pleistocene and Recent in age. Vents surrounding the cratered summit of Mt. Edziza have erupted within the past 1,200 years.

Quarried phyllite forms the foundation material for the ferry terminal at Wrangell. Glacially transported granite boulders lie along the beach. —CC

Phyllite of the Taku terrane crops out just north of the Wrangell ferry terminal. —CC

Wrangell Garnets

The ferry dock at Wrangell is built on quarried phyllite of the Taku terrane. Along the shoreline adjacent to the terminal, granitic boulders lie on top of the phyllite, probably left there by melting glacial ice. Visitors are likely to encounter young Wrangellians bearing muffin tins filled with garnets for sale. The stones come from Garnet Ledge, northeast of Wrangell near the Stikine delta, opposite the southern end of Sergief Island. In 1915, Alaska Garnet and Manufacturing Co. operated a mine there. This was the first corporation in the world composed entirely of women. The best stones were used for gems, the waste material for foundry powder. The ledge is now the property of Wrangell children, all adults need special permits to remove garnets.

Chief Shakes Hotsprings

Up the Stikine River about 12 miles is Chief Shakes Hotsprings on Ketili Creek. Chief Shakes was "the youngest but headest" of the Stickine Indian Chiefs according to John Muir and S. Hall Young. The hot springs lie about 15 miles west of the contact between layered gneiss associated with the Tracy Arm terrane and the metamorphosed phyllitic schist of the Taku terrane. Rocks near the spring are heavily fractured and the springs probably result from deep circulation of meteoric waters along fractures and faults. The temperature of the spring is 122 degrees F.

Stikine Icefield

Across the crest of the Coast Mountains from the Stikine River, approximately 120 miles to the Whiting River lies the Stikine Icefield. This large region of ice contains more than a dozen glaciers with lengths of ten miles or more. Many of the larger glaciers originate at elevations of more than a mile and flow down to sea level, where they terminate in the fjords of Frederick Sound and Stephens Passage.

Just north of the Stikine River delta is LeConte Glacier in LeConte Bay. This is the southernmost glacier reaching Frederick Sound and also the southernmost active tidewater glacier in North America. The enormous volume of icebergs produced by this glacier attracted San Francisco ice ships as early as 1853. LeConte glacier has retreated about two and one-half miles since its position was first charted in 1887, but has stabilized in the past two decades. John Muir visited the glacier in 1879 and described it as one of the most imposing and first class glaciers he had ever seen.

Fort Wrangell in 1898 featured totem poles, stores, tents, houses, boardwalks and Cassiar-bound miners. —U.S. Geological Survey photo by A.H. Brooks

Wrangell Narrows passes through rocks of the Gravina belt and poses a challenge to marine navigation. —Rod Flynn photo

Wrangell—Petersburg
Sailing time: approx. 3 hours
32 air miles

Wrangell Narrows

Most of the Alaska State Ferries follow the narrow passage between Mitkof and Woewodski Island during the early hours of the morning. Fortunately it is never very dark at night during the summer in Alaska so if one can stay awake, it is well worth the late night vigil to watch the ship's progress through the Wrangell Narrows. From the town of Wrangell it takes the ferry approximately 1 hour and fifteen minutes to turn around and reach the entrance to Wrangell Narrows. The 21 mile-long channel averages a half-mile in width but at its narrowest is only 100 yards wide. More than 70 channel markers, most with fixed or flashing lights, give the Narrows the appearance of a Christmas tree. Channel markers are green to the left and red to the right for ships sailing north. An outgoing tide adds a dramatic flavor to the voyage.

Woewodski Island forms the western shore of Wrangell Narrows at its southerly entrance. The Olympic Gold Mine is on the northwest side of Woewodski Island in rocks of the Alexander terrane. Much of Woewodski Island and all of Wrangell Narrows are in the Gravina belt. On central and eastern Woewodski Island are layers and lenses of the metallic minerals pyrite, pyrrhotite, and smaller amounts of arsenopyrite, sphalerite, galena, and chalcopyrite that are interbedded with metamorphosed volcanic and sedimentary rocks. The

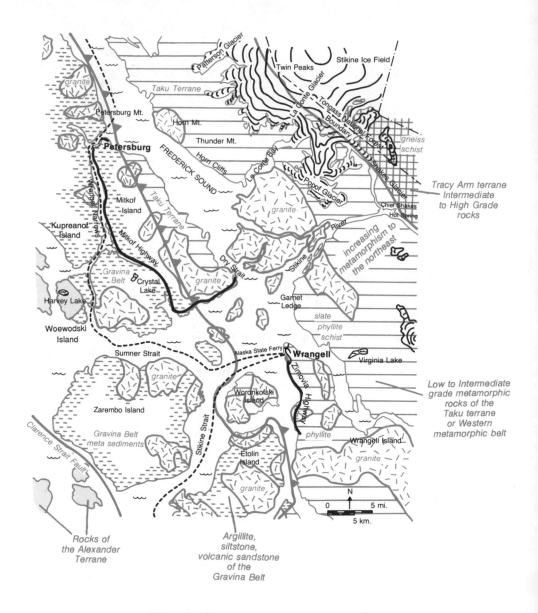

WRANGELL—PETERSBURG
Ferry Route

Petersburg area mineral deposits and their host rocks are part of a 300 mile long belt of similar deposits and host rocks that stretches the length of southeast Alaska from Juneau to Ketchikan. This belt contains volcanically generated massive sulfide and related barite deposits in metamorphosed late Triassic sedimentary rocks. In 1973, a barite mine was in operation on Big Castle Island in Duncan Canal northwest of Woewodski Island. Barite is used for paint pigment, barium salts, heavy drilling mud, and the manufacture of rubber, wallpaper, asbestos, cement, and enamel.

Petersburg

The ferry arrives in Petersburg approximately 3 hours after leaving Wrangell. North across Wrangell Narrows from Petersburg on Kupreanof Island is Petersburg Mountain at 2750 feet. This is one of many granite bodies that lie in a northwest-trending belt. These granite intrusions are part of the Admirality-Revillagigedo plutonic belt, which extends from Ketchikan to Juneau, outboard or west of the Coast Mountains. The central Lindenburg Peninsula and Horn Mountain on the mainland contain bodies of granitic rocks, relatively resistant stocks of varying size that rise steeply above the surrounding bedrock. The bedrock here consists of the Gravina belt. Exposures are poor and outcroppings are generally found either on nearly vertical mountain slopes above tree line, or along the intertidal zone on the coast. Some of the intrusions created contact aureoles which locally contain the mineral andalusite. Four granite bodies on Mitkof Island are similar in their mineral compositions but have different textures. One of those yielded an age date of 89.1 million years, late Cretaceous.

Petersburg located at the northeastern end of Wrangell Narrows on Mitkof Island. Granitic Petersburg Mountain lies across Wrangell Narrows on Kupreanof Island in the background. —Rod Flynn photo

A hike to the top of Petersburg Mountain provides excellent views to the east of mainland mountains with their glaciers. Devil's Thumb at 9077 feet and Kate's Needle at 10,102 feet are on the border between Alaska and British Columbia. Icebergs from LeConte Glacier drift north in Frederick Sound, toward the northern end of Wrangell Narrows, and sometimes beach under Horn Cliffs and Horn Mountain, 2960 feet, on the mainland opposite the northern tip of Mitkof Island. On rare occasion the icebergs land in Petersburg. A thrust fault that trends northwest through central Mitkof Island marks the zone of ovethrusting of rocks from the Gravina belt eastward onto rocks of the Taku terrane. The Mitkof highway, the major road on the island, leads 34 miles south from Petersburg to the Stikine River delta at the south end of the island. It provides excellent views of Wrangell Narrrows.

Recent Deposits in the Petersburg Area

In 1880 the main high waterline was along the present course of Main Street near the vicinity of the City Harbor. The area west of Main Street is mostly artificial fill quarried from local source areas. Southwest of town the airport runway is built up from as much as 27 feet of broken graywacke bedrock placed as fill over spongy peat deposits.

The organic deposits covering much of the surface are called muskeg or peat. They are made of plant materials such as sphagnum moss, sedges, heaths, and other small woody plants. The thickness of the peat deposits ranges from about 7 to 22 feet in this area.

Muskeg and peat deposits develop where the climate is cool and moist and where subsurface drainage is generally poor. The rate of accumulation of peat varies but may be about 1 foot per thousand years. Peat in the Petersburg region has been accumulating for the last 8,000 years.

The high porosity and high moisture content of peat make it a poor foundation material to build on. The amount of compression possible depends upon the proportion of wood fragments in the peat. The shear strength of peat is variable, but tends to be low. The vibration of heavy equipment may cause peat to lose its coherence and approach a jello-like consistency during heavy construction. In Prince Rupert, bulldozers were used to turn a peat mass 9 feet high by 225 feet long into a liquid pulp which could then be removed to expose firmer substrates below.

Where peat is more than 9 feet thick, foundations are usually set on pilings. Fill can generally be placed over peat in such a way that it floats uniformly without forcing the peat to flow out from underneath. At Ideal Cove, 26 miles south of Petersburg, on the east side of Mitkof Island opposite LeConte Bay on the mainland, the Forest Service successfully covered peat deposits with urethane foam, then with road fill.

Typically the peat deposits cover a layer of glacial sediment that was dumped from floating icebergs and sea ice, and bulldozed by tidewater glaciers. Later the mixture of silt, sand, and gravel was modified by underwater slumping and siding. Most of the modern shore area is underlain by these glacial-marine deposits at depths of 4 feet or so; they reach a thickness of approximately 9 feet.

Ferries sailing northward to Juneau enter Stephens passage at Cape Fanshaw following the east side of Admiralty Island, passing the entrance to Seymour Canal and the Glass Peninsula to the west, and Endicott and Tracy Arms to the east on the mainland.

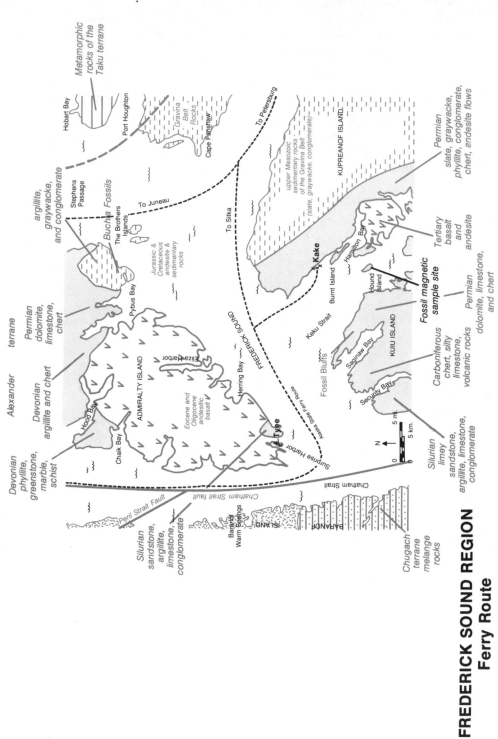

FREDERICK SOUND REGION
Ferry Route

Metamorphic rocks of the Taku terrane

Hobart Bay

Port Houghton

Cape Fanshaw

Gravina Belt Rocks

To Petersburg

upper Mesozoic sedimentary rocks of the Gravina Belt — (slate, graywacke, conglomerate)

KUPREANOF ISLAND

Permian slate, graywacke, phyllite, conglomerate, chert, andesite flows

Stephens Passage

To Juneau

The Brothers Islands

Bucha Fossils

argillite, graywacke, and conglomerate

Jurassic & Cretaceous andesite & sedimentary rocks

To Sitka

Kake

Hamilton Bay

Tertiary basalt and andesite

Pybus Bay

Permian dolomite, limestone, chert

Burnt Island

Hound Island

Fossil magnetic sample site

Permian dolomite, limestone, and chert

Eliza Harbor

ADMIRALTY ISLAND

FREDERICK SOUND

Keku Strait

Alexander terrane

Devonian argillite and chert

Herring Bay

Saginaw Bay

KUIU ISLAND

Carboniferous chert, silty limestone, volcanic rocks

Hood Bay

Eocene and Oligocene andesitic basalt

Fossil Bluffs

Chaik Bay

Devonian phyllite, greenstone, marble, schist

Tyee

Surprise Harbor

Alaska State Ferries Ferry Route

Security Bay

Silurian limey sandstone, argillite, limestone, conglomerate

5 mi.

5 km.

N

Chatham Strait

Peril Strait Fault

Chatham Strait fault

Silurian sandstone, argillite, limestone, conglomerate

Baranof Warm Springs

BARANOF ISLAND

Chugach terrane melange rocks

46

Petersburg—Sitka
Sailing time: about 10 hours
75 air miles approx.

Leaving Petersburg the ferry passes northeastward out of Wrangell Narrows and enters Frederick Sound. From this point it is 181 miles to Sitka, 123 to Juneau. On the mainland, Horn Mountain rises 2960 feet from sea level. The ferry next sails by the northwestern portion of Kupreanof Island and passes Thomas Bay on the mainland. Thomas Bay is the purported residence of Alaska's Big Sound, passing between Kupreanof and Kuiu Islands to the south and Admiralty Island to the north. Southwest of the southern tip of Admiralty Island, on northwestern Kupreanof Island, lies the community of Kake in Keku Strait. To the east are the marine mudstones and graywackes of the Gravina Belt.

The rocks on northwestern Kupreanof Island are part of the Alexander terrane. They are a Paleozoic sequence of metamorphosed basalt and limestone that overlies chert and tuff. The chert is a marine rock made up of microscopic particles of silica, tiny fossils of the marine plankton radiolaria. Tuff is a rock made up of compacted volcanic fragments. Studies of paleomagnetism, ancient magnetism trapped in iron minerals, of the Hound Island Volcanics in the Keku Islets area have shown that the Alexander terrane was far to the south during the early Paleozoic but that it had moved northward to about Latitude 19° North (presently the latitude of Mazatlan, Mexico) by late Triassic time. A latitude of about 24° North by the mid Cretaceous time is recorded in the fossil magnetism of the rocks near Point Camden, Keku Islets, and Kuiu Islands.

Four hours from Petersburg, approximately 103 miles to Sitka, the ferry passes by Surprise Harbor on the southwest tip of Admiralty Island.

Chatham Strait Fault

The ferry next turns north up Chatham Strait, following the Chatham Strait fault, long recognized as an important structural feature in southeastern Alaska. Geologists estimate that formerly adjacent rocks on either side of the fault have been separated by fault movement of about 90 miles. Offset of rocks across the fault is to the right just like the displacement along the San Andreas Fault in California, where Los Angeles on the west side of the fault is moving northward relative to San Francisco on the east side of the fault. In tens of millions of years the Los Angeles terrane may find its way to southeastern Alaska. Hollywood will never be the same.

Studies of movement along the Chatham Strait fault have been complicated by rocks offset vertically as well as horizontally. The water covering the fault in Chatham Strait and Lynn Canal hides other clues that might reveal the history of movement along this fault.

Rocks that have been separated by the Chatham Strait fault include a region of metamorphic rocks exposed west of the fault and west of Lynn Canal for 30 miles south from Haines. These rocks match parts of the Retreat group and the Gambier Bay formation east of the fault on southern Admiralty Island, about 89 miles away. Silurian sedimentary rocks called the Bay of Pillars formation have outcrops east of the fault on the west side of Kuiu Island and were probably deposited near the

Bands of chert in Paleozoic limestone at Security Bay, Kuiu Island.
—George Reifenstein photo

Point Augusta formation, presently exposed on northern Chichagof Island, opposite Hawk Inlet on Admiralty Island. These rocks represent respectively the inner and mid-fan sedimentary environments of an undersea turbidite sequence, and they have been offset by the Chatham Strait Fault.

Cretaceous sedimentary rocks of the Chugach terrane make up much of Baranof Island west of the Chatham Strait fault but do not crop out onshore east of it. The same rocks may be underwater south of the Hazy Islands and Coronation Island, southwest of Kuiu Island. A scuba-diving geologist would help here. On southern Admiralty Island, Eocene and Oligocene basalts lie next to the fault on the east side. Matching rock units may be underwater on the west side of the fault in Icy Strait, where dredged rock samples of the same age and rock type have been recovered. If this is a match, the basalts have been separated roughly 60 miles over the past 22.5 million years.

The northern end of the Chatham Strait fault extends onshore through the Chilkat River Valley near Haines. Studies there show that layered rocks deposited on the continent between 65 and 35 million years ago have been intensely deformed as a result of large-scale right-handed shear along the Chatham Strait fault, much like the movement of playing cards in a pack. This movement occurred sometime between 30 million and 10,000 years ago. No offset in Recent features onshore or in deposits filling the northern Lynn Canal fjords have been detected by ships studying bottom features using seismic reflection profiling.

Eastern Baranof Island

Five hours and 15 minutes from Petersburg, the ferry reaches Kelp Bay on eastern Baranof Island. Rocks here are part of the Chugach terrane and are equivalent to the Pinnacle Peak phyllite and the Waterfall greenstone on western Chichagof Island. The Goon Dip greenstone of western Chichagof Island does not occur in Kelp Bay. Hood Bay can now be seen on Admiralty Island east of the ferry.

Five hours and 50 minutes after leaving Petersburg, the ferry nears Peril Strait, a fault-generated waterway that separates Baranof Island to the south from Chichagof Island to the north. To the east across Chatham Strait on the western coast of Admiralty Island is the settlement of Angoon at the entrance to Kootznahoo Inlet. A chain of lakes and portage trails connect the head of Kootznahoo Inlet with Mitchell Bay on the east coast of Admiralty Island, making this a popular canoeing route across the island.

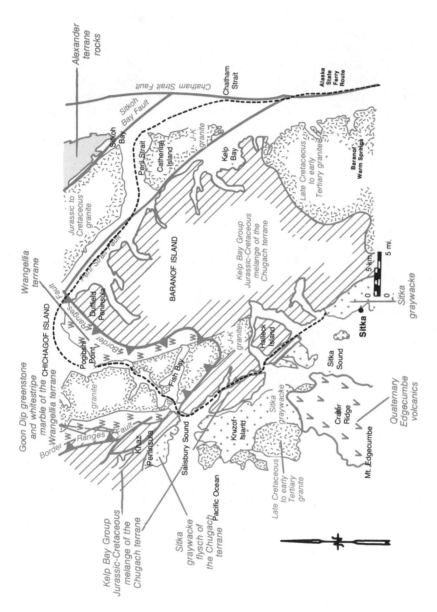

PERIL STRAIT–SITKA
Ferry Route

Alexander terrane rocks

Chatham Strait Fault

Chatham Strait

Sitkoh Bay Fault

Alaska State Ferry Route

Sitkoh Bay

Peril Strait

Catherine Island

Kelp Bay

Jurassic to Cretaceous granite

J-K granite

Late Cretaceous to early Tertiary granite

Baranof Warm Springs

Wrangellia terrane

Peril Strait Fault

BARANOF ISLAND

Kelp Bay Group Jurassic-Cretaceous melange of the Chugach terrane

Border Ranges Fault

Duffield Peninsula

Pogibshi Point

Fish Bay

Halleck Island

Sitka Sound

Sitka

5 mi.

5 km

0

0

Sitka graywacke

J-K granite

Goon Dip greenstone and whitestripe marble of the CHICHAGOF ISLAND Wrangellia terrane

granite

Border Ranges Fault

Khaz Peninsula

Salisbury Sound

Kruzof Island

Sitka graywacke

Crater Ridge

Mt. Edgecumbe

Quaternary Edgecumbe volcanics

Kelp Bay Group Jurassic-Cretaceous melange of the Chugach terrane

Sitka graywacke flysch of the Chugach terrane

Pacific Ocean

Late Cretaceous to early Tertiary granite

Peril Strait and its Faults

The ferry now turns west into Peril Strait. The northwest-southeast fault that formed this waterway can be traced 92 miles from the entrance of Peril Strait, northwest through Hoonah Sound, and Lisianski Inlet. The fault can be seen on Catherine Island at the entrance to Peril Strait and between Hoonah Sound and Lisianski Inlet where it forms a vertical zone of broken rock as much as a mile wide. Rocks across the fault have moved to the right as in the Chatham Strait fault, and an offset of about 19 miles can be seen in rocks adjacent to the fault on northwestern Chichagof Island. Movement seems to have occurred before the intrusion of igneous plutonic rocks in Tertiary time.

Rocks on the north side of Peril Strait, on southern Chichagof Island, are part of the Alexander terrane, similar to rocks on Prince of Wales Island. They were intruded by younger Tertiary igneous rocks. Along the northern shoreline near the entrance to Peril Strait is the entrance to Sitkoh Bay. Ash that erupted from Mt. Edgecumbe about 9,000 years ago can be found here. Due north of Peril Strait in the next drainage is the Kadashan River Valley on Tenakee Inlet opposite Tenakee Springs.

The broad swaths of treeless land on either side of Peril Strait are not newly deglaciated surfaces but rather lands that have been clear-cut by local loggers. The logs went to a pulp mill in Sitka.

Northern Baranof Island is made up of rocks of the Chugach terrane. These complexly deformed and metamorphosed upper Mesozoic graywackes, shales, and volcanic rocks were scraped from the sea-floor onto the continent as the Chugach terrane was accreted.

Seven hours and twenty-five minutes from Petersburg the ferry sails around the Duffield Peninsula and Peril Strait to the southwest toward Sitka. Hoonah Sound branches off to the northwest following the Peril Strait fault. The ship is now following the southern boundary of the Border Ranges fault in southeastern Alaska. Here a piece of the Wrangellia terrane, the rocks now exposed north of us on Chichagof Island, was sliced off and sandwiched between the Alexander terrane to the northeast and the Chugach terrane to the southwest.

About eight hours and 45 minutes from Petersburg the ferry sails around Pogibshi Point and enters a narrow southwesterly trending section of Peril Strait. At low tide this passage keeps ferry captains on their toes as they carefully maneuver their ships to avoid dangerous shoals. As the ferry enters Sergius Narrows red channel buoys are kept on the left, green markers on the right. During tidal changes the water seems to rip over the shallow reefs and bars. Next the ferry leaves Sergius Narrows and enters Salisbury Sound. The course is now to the south as the ship heads toward Sitka. The safe shelter of the Inland Passage is behind us and the open water of the Pacific Ocean lies off to the west. The Conical Peaks on the Khaz Peninsula to the north on Chichagof Island, are rocks of the Chugach terrane.

Northwestern Chichagof Island

The sedimentary and metamorphic rocks of western Chichagof and Yakobi islands form four northwest-southeast linear belts. The oldest rocks on the northeast form a discontinuous belt along Hoonah Sound and Lisianski Inlet. They are Mesozoic and Paleozoic sedimentary and volcanic rocks that have experienced medium- to high-grade metamorphism. Younger granitic rocks that intrude them have been sheared by movement along the Peril Strait fault.

The next belt of rocks to the southwest is composed of the White-stripe marble and the Goon Dip greenstone. They were once thought to be part of the Kelp Bay group and the Chugach terrane. Their striking resemblence to the Nikolai greenstone and Chitistone limestone in the Wrangell Mountains suggests they came to Alaska as part of the Wrangellia terrane. Both of these northeastern belts are very disrupted by igneous intrusions emplaced during Jurassic and Cretaceous time.

The two northeastern belts are separated from two belts of younger rocks to the southwest by the Border Ranges fault, which separates similar belts of rocks in the Chugach Mountains in south central Alaska. This fault is the record of a mid-Cretaceous to early Tertiary accretionary event that placed the Cretaceous Chugach terrane rocks to the west against the Wrangellia terrane to the east. An accretionary event is to geologists as a nine car pile-up on a major highway is to an insurance adjuster; each must sort out which pieces belong to which original terrane or automobile.

The Kelp Bay group makes up the third belt in the area. It consists of a tectonic patchwork zone of many different late Mesozoic metamorphosed sedimentary and volcanic rocks that include greenschist, phyllite, greenstone, tuff, graywacke, mudstone, marble, and chert. These units typically occur as very deformed, irregular, fault-bounded blocks that, because of their extremely disrupted nature, are considered to be a melange within the Chugach terrane. Melange is a word borrowed from the French meaning mixture; it is used by geologists to describe the chaotic assortment of seafloor sediments and volcanic rocks that jumble together when terranes slowly crash into continents.

Southwest of the Kelp Bay group is the youngest and fourth belt of rocks, the Sitka graywacke. This is a broken formation composed of Cretaceous sandstones, siltstones, and conglomerates that formed near the edge of the continental shelf. Sitka graywacke crops out especially well near the town of Sitka. These rocks closely resemble graywackes found in Kelp Bay on eastern Baranof Island, in the Valdez formation in Prince William Sound, and in the Shumagin formation south of the Alaska Peninsula. All of these graywackes are dirty sandstones that are grayish-green in color. They formed in thick sequences of sand and mud as volcanically derived sediments shed into the Gulf of Alaska during Cretaceous time. As the Chugach terrane docked against the southern and southeastern Alaskan coast these sediments were plastered onto the continent.

Mount Edgecumbe

The ferry next passes Mount Edgecumbe on Kruzof Island. In late Pleistocene and recent time a northeast-trending line of volcanic vents across the island erupted to produce the Mount Edgecumbe volcanic field, which covers about 100 square miles. These very young volcanic rocks on Kruzof Island include flows of basalt, basaltic andesite, andesite, and lighter rocks, as well as rhyolite domes. A large composite cone, Mount Edgecumbe, and a collapsed caldera, Crater Ridge, dominate the island; smaller cinder cones are scattered throughout. A prominent ash layer, found as far northeast as Juneau, erupted from the Mount Edgecumbe field about 9,000 years ago. The volcano has been inactive during the past 200 years. A Sitka resident celebrated April Fool's Day of 1974 with an old tire roast in the caldera at the top of Mount Edgecumbe.

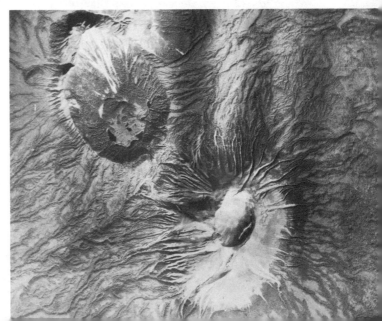

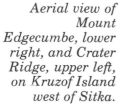

Aerial view of Mount Edgecumbe, lower right, and Crater Ridge, upper left, on Kruzof Island west of Sitka.
—G.L. Snyder photo

53

The Mount Edgecumbe basalts are relatively rich in silica and aluminum, low in such elements as sodium, potassium, and titanium. Such rocks originate deep in the earth's mantle and closely resemble the basalts of the Hawaiian Islands, which formed in the middle of the Pacific Ocean, far from the edge of the continent. Basalt lavas of this type are very fluid and flow easily across the countryside to form gently sloped shield volcanoes. They are very different from the volcanic rock that formed the Cascade vocanoes like Mt. Rainier and Mt. Saint Helens.

Why did Mt. Edgecumbe form here near Sitka along the continental margin of southeast Alaska far from the Aleutian trench? The nearest volcanic rocks of similar age are 150 miles away to the east in British Columbia. The Edgecumbe volcanic field lies near the boundary between the Pacific and North American plates although there are not many earthquakes in this region to help geologists precisely locate the plate junction. The aftershock zones of major earthquakes along the Fairweather fault to the northwest of Sitka near Yakutat, suggest that Mt. Edgecumbe is within 18 miles of the plate boundary on the North American side. Many geologists believe that the Queen Charlotte transform fault is leaking basalt magma that makes its way to the surface to form Mt. Edgecumbe. Local tectonic structures such as the Chatham Strait fault and other northwest-trending faults became inactive several tens of million years before the Edgecumbe volcanic rock erupted and thus cannot be involved.

The Mt. Edgecumbe volcanic field lies within a zone of Tertiary granitic plutons scattered over a region 38 miles by 73 miles. Chunks of this older granitic rock were carried up the throat of Mt. Edgecumbe by the younger lavas and ejected. These lavas flowed over the 100 million year old Sitka graywacke.

Walls of Sitka graywacke support an old Russian cannon on Castle Hill in downtown Sitka. —cc

Sitka Walking Geology Tour

About ten hours after leaving Petersburg the ferry docks at the ferry terminal 6 miles north of Sitka. Nearby is Stargaven Creek, which enters the ferry terminal bay. A fish camp at the creek mouth was used in the summer time by the Kitsadi people. The camp is sometimes called Old Sitka. The Kitsadi were the "Frog People" who canoed up from Nass River region in British Columbia by way of Wrangell. After slaughterous dealings with the Russians they relocated in Sitkoh Bay near the east entrance to Peril Straits.

A short ride into town provides roadside geologists with an opportunity to see many outcrops of Sitka graywacke, especially where construction workers have excavated down to bedrock. Notice the soils lying on top of the Sitka graywacke in the roadcuts enroute to town. A prominent 1-2" band of 9000 year-old ash from Mt. Edgecumbe is well preserved. In the downtown area be sure to visit Castle Hill, a prominent knob of graywacke that was favored by the Tlingits and later by the Russians for settlement because of its commanding view of the neighborhood. From the top of the hill you have excellent views of Mt. Edgecumbe, downtown Sitka, and Japonski Island. The wall supporting the old Russian cannons is made out of Sitka graywacke. This grayish sandstone is poorly sorted and has both fine and coarse grains. It is interbedded with thin to medium layers of hard mudstone and was subjected to low-grade metamorphism. Original sedimentary structures can be seen if you look closely. Near the mouth of Sawmill Creek, south of Castle Hill, the graywacke is overlain by volcanic ash. The Sitka graywacke is thought to be Cretaceous because outcrops on Kruzof Island contain marine invertebrate fossils of that age.

From Castle Hill, walk toward the crescent-shaped harbor and the Sheldon Jackson College and Museum complex. Graywacke outcrops can be seen throughout Sitka and along the Indian River. A return to the downtown area from the Indian Fort site along the beach at low tide provides an opportunity to collect some cherty rocks amongst the graywacke cobbles. Nice views of Mt. Edgecumbe and Mt. Verstovia make this a pleasant walk. Downtown there are two boulders of recently erupted Mt. Edgecumbe volcanic rock in the planter in front of the bank near the taxi stand. The holes in these rocks were made by escaping gas bubbles while the lava was still molten.

Sitka—Juneau
Sailing time: about 8 hours
95 air miles

The ferry retraces its route northward from Sitka, reentering Peril Strait and sailing eastward toward Chatham Strait between Chichagof and Admiralty islands.

At the eastern end of Peril Strait, the ferry turns north up Chatham Strait between Chichagof Island to the west and Admiralty Island to the east. Rocks along this leg of the trip are part of the Alexander terrane.

Tenakee Fault System

North of Peril Strait is a parallel waterway called Tenakee Inlet, well known as the site of Tenakee Springs. This region contains the Tenakee fault system, a complex of northwest and north-northwest-trending faults that cut through Chichagof Island northeast of Peril Strait, Hoonah Sound, and Lisianski Inlet. Included in the fault complex are the Freshwater Bay, Indian River, and Sitkoh Bay faults. These have right-lateral displacement of a few miles and vertical displacement of about a half mile. Tenakee Hot Springs is supplied by ground water moving through faults where it is heated at depth and then recirculated back to the surface.

Upward movement on the southwest side of the Tenakee fault system was caused by uplift of the Chichagof Island pluton sometime after about 130 million years ago. The Tenakee fault system probably extends to the northwest into Cross Sound and joins the fault systems on the west side of the main arm of Glacier Bay. The uplifted block of plutonic rocks on Chichagof Island continues northwest across Cross Sound into the Fairweather Range fault block, now the highest plutonic rocks in southeastern Alaska.

The ferry next passes Freshwater Bay, the site of a logging camp. Just north of Freshwater Bay, on Inyoukeen Cove, is the abandoned mining town of Gypsum. The gypsum mine that once operated here used one mile of track and a locomotive to deliver the rock to the wharves that extended 2,000 feet to deep water. The gypsum went to a mill in Tacoma, Washington where it was used for wall plaster, fertilizer, and cement.

The ferry continues north up Chatham Strait, traversing the western coast of Admiralty Island. As the ferry passes Point Augusta on the northeast corner of Chichagof Island, the narrow mouth of Hawk Inlet, on Admiralty Island, appears off the eastern side of the ship.

Greens Creek Gold Mines, Northwest Admiralty Island

Eight miles up Greens Creek on the south side of Hawk Inlet a large deposit of silver, lead, zinc, gold, and copper sits in rocks of the Alexander terrane. The first clue to the discovery came in 1974 when geologists discovered abnormally high assay values in panned sediments from the Greens Creek delta. Later, with the help of a helicopter, a reddish stained rock outcrop was spotted at an elevation of 1700 feet that led geologists to the main ore body. Drilling revealed the existence of an ore body 100 feet wide and more than 1900 feet deep—over 3.5 million tons of recoverable ore reserves or about 20 years of mining at a rate of 85,000 tons-per-year.

The ore deposits of Greens Creek may have been formed in a deep marine basin behind an active volcanic island chain. Hot vents on the basin floor may have transferred the ore minerals into the marine sediments. A clam of Triassic age, preserved in a clay nodule, was protected from the folding and faulting that occurred next and remained in the rock for more than 250 million years until it was discovered by Greens Creek miners.

The Greens Creek Mine opened in February 1989 and began full production that year. The largest source of silver in North America was forced to suspend mining and shut down the operation in May of 1993 due to a decline in silver prices from $11.58/oz. in 1989 to $3.70/oz. in 1993.

Chilkat Peninsula

The southern tip of Chilkat Peninsula next appears off the west side of the ship. Rocks on the southern Chilkat Peninsula have been mapped as a Silurian sequence equivalent to the Bay of Pillars formation on Kuiu Island in the southern portion of the Alexander terrane. Here the rocks are known as the Point Augusta formation; they are offset from their southerly equivalents by about 90 miles along the Chatham Strait fault, which the ferry has been following. The rocks are sandstones composed mainly of broken igneous rock fragments rather than quartz. Also present in the sandstone are grains of volcanic rocks that tell of active volcanism in the areas that were supplying the sediments.

Northern Admiralty Island

North of Hawk Inlet is Funter Bay, the proposed site for an Alaska Marine Park. An intrusion of granitic rock into the Alexander terrane rocks crop out along the northwestern tip of Admiralty Island. As the ferry rounds the northern tip of the Island and traverses Barlow Cove, rocks of the Alexander terrane are left behind to the west and we reenter the Gravina Belt. This portion of the belt contains the Douglas Island volcanic rocks, dark green andesite flow breccias and tuff with crystals of augite and some hornblende as much as 1/2 inch across. They have been mapped mainly on the Glass Peninsula of Admiralty Island, south of Douglas Island. No fossils have been found in these rocks, but geologic relationships with other rocks makes it seem likely that they formed during Jurassic and Cretaceous time. This leg of the trip ends at Auke Bay, the ferry terminal for Juneau, about 14 miles north of the capital city. The northern end of Douglas Island appears off the south side of the ferry before it docks, and in good weather a portion of Mendenhall Glacier, called the "Auk Glacier" by John Muir, appears nestled in the mountains east of Auke Bay.

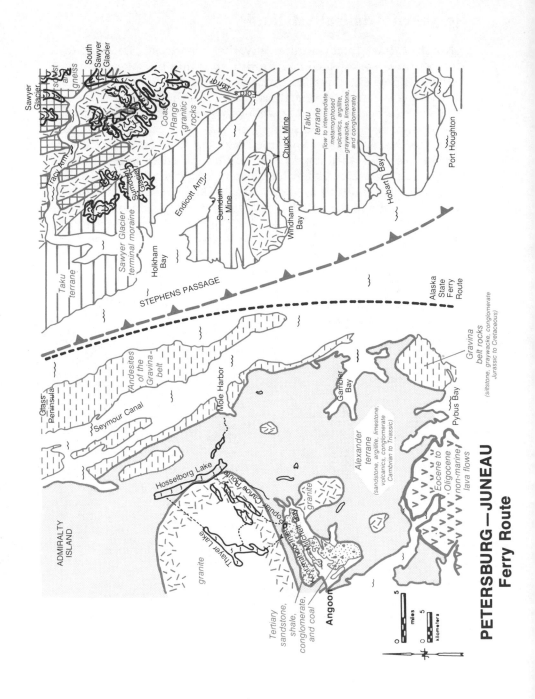

Sawyer Glacier

South Sawyer Glacier

schist and gneiss

Coast Range granitic rocks

Taku terrane
(low to intermediate metamorphosed volcanics, argillite, graywacke, limestone, and conglomerate)

Fords Terror

Chuck Mine

Port Houghton

Hobart Bay

Tracy Arm

Sumdum Glacier

Endicott Arm

Sumdum Mine

Windham Bay

Sawyer Glacier terminal moraine

Holkham Bay

Taku terrane

STEPHENS PASSAGE

Alaska State Ferry Route

ADMIRALTY ISLAND

Glass Peninsula

Andesites of the Gravina-belt

Seymour Canal

Mole Harbor

Gambier Bay

Pybus Bay

Gravina belt rocks
(siltstone, graywacke, conglomerate Jurassic to Cretaceous)

Alexander terrane
(sandstone, argillite, limestone, volcanics, conglomerate Cambrian to Triassic)

Hosselborg Lake

granite

Cross Admiralty Canoe Route

Thayer Lake

Mitchell Bay

Kootznahoo Inlet

granite

Eocene to Oligocene non-marine lava flows

Tertiary sandstone, shale, conglomerate, and coal

Angoon

miles

5

0

kilometers

5

0

N

PETERSBURG—JUNEAU
Ferry Route

Petersburg—Juneau
Sailing time: about 7½ hours
120 air miles

From Petersburg the ferry travels north out of Wrangell Narrows and into Frederick Sound, between the northern end of Kupreanof Island to the southwest and Thomas Bay on the mainland to the northeast. At Cape Fanshaw the ferry heads north into Stephens Passage toward Juneau.

Stephens Passage Group

Off the west side of the ferry are three assemblages of rocks that are part of the Gravina Belt. This narrow belt of middle Jurassic to middle Cretaceous andesites, graywackes, other marine sedimentary rocks, and conglomerate lies upon Paleozoic rocks of the Alexander terrane. The Gravina Belt rocks were probably deposited as marine sandstones in a chain of volcanic islands as the Alexander terrane was adding itself to the North American continent. The Gravina Belt, especially on Gravina Island near Ketchikan, is deformed and metamorphosed, probably the result of northeast movement of the Alexander terrane during docking. Rocks of this belt include the Seymour Canal formation, made up of upper Mesozoic sedimentary rocks, and the overlying Douglas Island and Brothers volcanic rocks.

Southern Admiralty Island

Two large islands known as The Brothers lie off the southeast coast of Admiralty Island just north of the entrance to Pybus Bay. The Brothers volcanic rocks are about 2,000 feet of well-bedded, dark andesite flows, broken and refused volcanic breccias, and some sedimentary rocks made up of volcanic fragments. No fossils have been found, but the rocks are thought to be Jurassic and Cretaceous in age.

Rocks on Admiralty Island near Pybus Bay are mudstones, graywacke sandstone, and conglomerates that have been faulted and severely folded but little metamorphosed. The mudstone contains fossils of the extinct bivalve or mussel-like animal known as *Buchia* that is found in rocks of Mesozoic age throughout Alaska. The rocks are in layers that total between 4,000 and 8,000 feet thick.

From Gambier Bay northward along the east side of Seymour Canal on eastern Admiralty Island, the Gravina belt rocks are folded on a regional scale. Here they consist of slate, phyllite, siltstone, graywacke, and conglomerate, but no fossils.

Mole Harbor north of Gambier Bay, on the west side of Seymour Canal, serves as the endpoint for a popular canoeing route across Admiralty Island. A series of well-placed lakes allows boaters with strong backs to canoe and portage from east to west, through the Mesozoic rocks of the Gravina belt, the Paleozoic rocks of the Alexander terrane, across a Cretaceous granitic pluton, and emerge on the west side of Admiralty Island into Mitchell Bay. Young Tertiary sandstones, shales, conglomerates, and even coal deposits there, surround Kootznahoo Inlet near the town of Angoon, alongside Chatham Strait in Paleozoic rocks of the Alexander terrane.

The Glass Peninsula, on the east side of Seymour Canal, was named by the U.S. Coast and Geodetic Survey for Commander Henry Glass who surveyed this area in 1881. The rocks here are dark green andesite flows; some composed of broken volcanic fragments; some are tuffs, made up of ash and volcanic particles that were air-borne during eruption. Crystals of augite and hornblende up to a half inch in diameter are found. The rocks have been metamorphosed to schist on the eastern Glass Peninsula. They are called the Douglas Island volcanic rocks after similar outcrops on Douglas Island near Juneau. These Douglas Island Volcanics and the Brother's Volcanics are Jurassic to Cretaceous in age.

Tracy Arm—Ford's Terror Wilderness Area

Stephens Passage follows a fault that separates rocks of the Alexander terrane and Gravina belt from the more metamorphosed rocks of the Taku and Tracy Arm terranes, which are part of the Coast Metamorphic and Plutonic complex to the east. This fault extends northward into Gastineau Channel between Juneau and Douglas Island.

Rocks exposed between Port Houghton and Juneau, and especially those in Endicott and Tracy Arms have been identified as part of the

Taku and Tracy Arm terranes. These rocks were metamorphosed between middle Cretaceous and Miocene time. They probably formed as a sort of tectonic and metamorphic "continental bruise" during the docking of the Wrangellia and Alexander terranes against the Stikine terrane. This occurred just before the Chugach terrane attached itself to the seaward side of Wrangellia. Geologists are uncertain of the identity of the rocks in the Taku and Tracy Arm terranes before they were metamorphosed. Evidence from fossils, structures in the rocks, and the rocks paleomagnetic history suggest that the Tracy Arm terrane could have been produced by cooking the lower section of the Alexander terrane and that the Taku terrane may be a metamorphosed portion of the upper Alexander or lower Stikine terrane rocks. Thus the Tracy Arm is probably not a legitimate terrane.

Geologists sometimes lump smaller rock units into larger groups once they have gathered enough information from splitting rocks into smaller and smaller units that they can at last see the big picture. One might compare it to the sorting of socks at the laundromat. Eventually large piles of separately paired socks end up in the same dresser drawer, unless of course they have vanished somewhere inside the drier. There will be a great deal of sorting and resorting of Alaskan terranes before geologists fully understand their histories.

Southern End of the Juneau Gold Belt

Gold was first discovered in 1889 at Sumdum and Windham Bay, in stream gravels washed down from ore bodies deep in the mountains. The mineralized belt stretches over 100 miles from Windham Bay to Juneau and 60 miles farther north to Berner's Bay. This string of deposits is known as the Juneau Gold Belt.

The Sumdum Mine produced nearly half a million dollars in gold and silver before it closed in 1903. The local Tlingit Indians created the name "Sumdum" to mimic the thundering echo of ice calving off glacier fronts into the deep fjords.

Tracy Arm Fjord

Icebergs at the entrance to Holkham Bay signal the presence of a tidewater glacier somewhere in the area. Tidewater glaciers can rapidly calve icebergs if they are in a retreating phase and recede from the shoal formed by the debris pile or moraine that they pushed up during times of advance. Glaciers grounded in deep water tend to lose ice rapidly and in large chunks. Once glaciers have melted back to shallow water, usually at the head of the fjord, the rate of retreat slows. On the southeast side of Harbor Island at Holkham Bay's entrance, two red vertical boards are used to line up the ship with the appropriate point on the mainland to avoid running aground on the Sawyer Glacier end moraine, which lies barely submerged at high tide across the entrance to Tracy Arm fjord. Once that tricky navigational maneuver is accomplished, a 25 mile, awe-inspiring trip through narrow walls of schist, gneiss, and granite awaits enthusiastic and, one hopes, warmly dressed travelers. The fjord is less than a mile wide for most its length and the rocky walls rise to elevations of 2,000 feet or more. Its greatest depth is 1,242 feet. The fjord formed as the North Sawyer and South Sawyer glaciers retreated. They were probably joined within the past 200 years, perhaps as recently as 1880. Their positions have fluctuated within this century but both have receded about 25 miles from their latest advance position down the fjord.

Newly calved ice near Sawyer Glacier in Tracy Arm fjord. —Rod Flynn photo

THE GREATER JUNEAU AREA

Metamorphic and Granite Rocks in the Juneau Region

The Juneau area provides many excellent exposures of metamorphic rocks. Across a belt from Douglas Island to the Juneau Icefield, parallel zones of slates, phyllites and schists, all metamorphosed marine shaley rocks, have been cooked and squeezed under high temperatures and pressures that increased to the east. Some of the best exposures are on the ridges northeast of the capital city. Gold occurs in quartz veins in these metamorphic rocks.

Before they were deeply buried and metamorphosed these Juneau rocks were shales, sandstones and volcanic rocks, probably part of the Alexander or Stikine terranes. During the docking of the Chugach terrane onto Wrangellia, the rocks of the inboard Alexander and Stikine terranes were deformed and heated by regional metamorphism. From eastern Douglas Island across Gastineau Channel to the mainland, there are exposures of greenschist, distinctive slates, phyllites, and schists that contain green minerals such as chlorite, actinolite, and epidote.

Metamorphosed siltstones and mudstones are on Blackerby Ridge, which separates Salmon and Lemon creeks about 5 miles north of Juneau. A hike along the ridge to the northeast provides outcrops of rocks bearing the index minerals biotite, garnet, staurolite, kyanite, and sillimanite. These minerals record an eastward increase in the temperature of metamorphism. At the easternmost end of the ridge, younger granitic rocks intruded the schist to form a migmatite composed of alternating layers of schist and granite.

The western part of the greenschist belt, on Douglas Island, contains rocks with minerals characteristic of lower temperature and lower pressure metamorphic conditions—albite, white mica, chlorite, pumpellyite, epidote, actinolite, and locally prehnite and stilpnomelane. These rocks were originally basaltic tuffs and breccias interlayered with graywacke and mudstone.

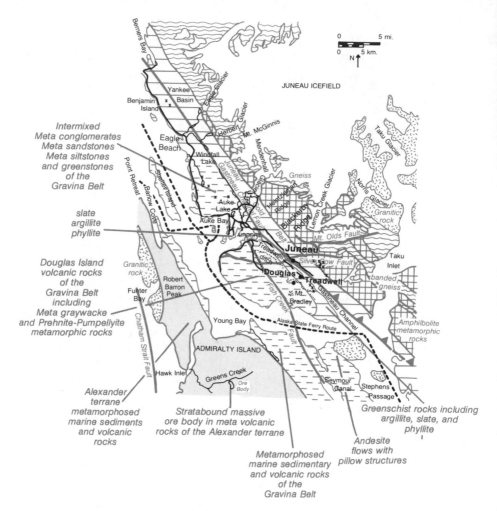

Intermixed
Meta conglomerates
Meta sandstones
Meta siltstones
and greenstones
of the
Gravina Belt

slate
argillite
phyllite

Douglas Island
volcanic rocks
of the
Gravina Belt
including
Meta graywacke
and Prehnite-Pumpellyite
metamorphic rocks

Alexander
terrane
metamorphosed
marine sediments
and volcanic
rocks

Stratabound massive
ore body in meta volcanic
rocks of the Alexander terrane

Metamorphosed
marine sedimentary
and volcanic rocks
of the
Gravina Belt

Andesite
flows with
pillow structures

Greenschist rocks including
argillite, slate, and
phyllite

JUNEAU ICEFIELD

0   5 mi.
0   5 km.
N

Berners Bay
Yankee Basin
Benjamin Island
Eagle Glacier
Herbert Glacier
Mt. McGinnis
Eagle Beach
Windfall Lake
Mendenhall Glacier
Juneau Gastineau Channel
Gneiss
Taku Glacier
Point Retreat
Shelter Island
Barlow Cove
Auke Lake
Auke Bay
Heintzleman Ridge
Blackerby Ridge
Lemon Creek Glacier
Norris Glacier
Granitic rock
Lemon
Gold Belt
Mt. Olds Fault
Juneau
Treadwell dike
Silverbow Fault
Taku Inlet
Granitic rock
Robert Barron Peak
Funter Bay
Douglas
Treadwell
banded gneiss
Mt. Bradley
Gastineau Channel
Young Bay
Fish Creek
Alaska State Ferry Route
Amphibolite metamorphic rocks
Chatham Strait Fault
ADMIRALTY ISLAND
Hawk Inlet
Greens Creek
Ore Body
Seymour Canal
Stephens Passage

JUNEAU REGION

Granitic rocks occur as many distinct plutons in the core of the coastal mountains northeast of Juneau. This is but one segment of an 8,000 mile-long rock complex that extends the length of the North American Cordillera from Baja California to the Aleutian Islands. Granitic intrusions mapped near Juneau include the Mount Juneau, Mendenhall Glacier, Lemon Creek Glacier, Carlson Creek, Dead Branch, and Annex Lakes plutons. This 70 to 55 million-year-old quartz-rich diorite is known as the Great Tonalite Sill and extends the length of southeast Alaska.

Downtown Juneau

The capital city of Alaska nestles at the base of Mount Juneau at 3,567 feet and Mount Roberts at 3,819 feet. From the summit of Mount Juneau, the Fairweather Range can be seen on sunny days about 150 miles to the northwest, lying west of Glacier Bay. Juneau is separated from Douglas Island to the southwest by the Gastineau Channel, which trends northwest along the Gastineau Channel fault. The fault continues northwest across the Mendenhall Peninsula, along the Montana Creek-Windfall Lake Valley, and across Berner's Bay. Eventually it intersects at the Chatham Strait fault in Lynn Canal south of Haines. The Gastineau fault does not show any recent offset and, like the Chatham Strait fault, has probably not moved since Tertiary time.

Good outcrops of greenschist metamorphic rocks can be seen on Telephone Hill, especially near the 5th Street entrance to the State Office Building. The Tokeen marble columns at the Capitol were quarried from marble deposits on Prince of Wales and adjacent islands. Glacial ice gouged the entrance to upper Gold Creek Valley, which can be seen especially well from the Douglas Bridge. Post glacial sands and gravels collected in a lake formed by a huge slab of rock that broke off and slid down into Gold Creek Valley from Mt. Juneau. Water stored between 75-200 feet in these sediments, is currently pumped from wells and used by the cities of Juneau and Douglas for drinking water.

Gold Mining in the Juneau Area

In 1897 John Muir and Presbyterian minister S. Hall Young paddled a large canoe south along the southeast Alaskan coast, returning to Wrangell after their visit to Glacier Bay and northern Lynn Canal. Muir was impressed with the beauty and splendor of Gastineau Channel and likened it to Yosemite Valley in California. He reported on the mineralized appearance of the mainland shore. Muir's information and rock samples brought by Chief Koweeh (Cowee), a sub-chief of the Auk Tlingits, from the Gastineau Channel area, made their way to George Pilz, a mining engineer in charge of the Stewart Claim at Silver Bay, southeast of Sitka. Pilz, seeing the gold, in 1880 contracted two "broke Cassiar miners," Joe Juneau and Richard Harris, to prospect for and locate mines in the Windham and Sumdum Bay region. The two, accompanied by three Indians who probably led them all the way to the gold, made their way north into Gastineau Channel and eventually followed Gold Creek, now in downtown Juneau, into Last Chance Basin near the present site of the Old Mining Museum. Then they went up Snowslide Gulch and over the summit of the mountain to find streaks and lumps of gold in the quartz of Silver Bow Basin.

Basin Road leads from 6th Street in downtown Juneau eastward up into the valley of Gold Creek. Initially the basin streams and gravels were worked for placer gold. In 1885 small claims in the Silver Bow Basin were consolidated for lode or hardrock mining. This group of claims became known as the Perseverance. A 10-stamp mill was completed in 1890, but a snowslide destroyed it 5 years later. A 100-stamp mill replaced it in 1907 but it was burned in 1912.

The hardrock miners worked quartz veins that stood more or less vertically in the dark slates, greenstones, and dark brown gabbro.

The Alaska Juneau Mill operated from 1917 – 1944.

—U.S. Geological Survey
photo by R.H. Sargent

By about 1915 miners were following quartz veins and ore bodies through schist. Frequent cave-ins of the schist caused waste rock above the quartz veins to fall into the newly mined high-grade ore. Since the success of the mining operation lay in hand sorting the ore rock from waste rock on the conveyor belts leading to the Alaska-Juneau Test mill, this made additional work for the milling crew.

The gold ores in Juneau and Douglas are part of the linear belt of gold deposits in northern southeast Alaska. In early Tertiary time, change in the direction of Pacific plate motion from north to northwestward, possibly due to India's collision with Asia, redirected the forces acting on the Gastineau Channel Fault from compressional to strike slip. The plate dynamics generated heat, caused fluids to migrate through rocks, and concentrated metal ores to create the Juneau Gold Belt. The Gastineau Channel Fault is a local name for the Coast Range Megalineament, a fault structure of great tectonic significance in this region.

Thane Road Log

Southeast of the capital city, beginning at the downtown ferry terminal is the road to Thane, once a separate and bustling mining town. At about 0.1 mile along the road is the Alaska-Juneau Mill on the hillside to the left. This mill handled 12,000 tons of ore per day. Dark waste rock was hauled by railcars to beachline where it was carried by conveyor belts to create a harbor structure for the Juneau waterfront. The city of Juneau also used waste rock for road building. The Perseverance Mines thus functioned as a giant rock quarry as well as a gold producer.

At mile 0.5, mine tailings from Alaska-Juneau Mill underlie oil storage tanks, the city's sewage treatment plant, and the former site of Juneau's "Million Dollar" Golf Course. At mile 2.3, a waterfall marks a major avalanche chute that can be deadly in the winter when snow accumulates on the upper slopes of Mount Roberts. From this point there are excellent views across Gastineau Channel to Douglas and the old Treadwell Mine Complex ruins at the south end of Douglas town.

Old Thane Mill Camp buildings survive near the beach at mile 3.5—remnants of a town that once boasted a meat market, bakery, warehouse, boarding house, general store, gymnasium, clubhouse, library, and two-room schoolhouse. The First World War claimed many mine workers, reducing the labor force. Much caving-in in the stope areas continued to mix waste rock with the ore. Run-off water flooded the stopes by 1921 making the ore too wet to mill, and the company ceased operation.

At mile 3.7, the Sheep Creek trail climbs 1000 feet to beautiful Sheep Creek valley, dominated by Sheep Mountain at 4,233 feet. Mining Camp ruins are all that remain of the old Portal Camp adjacent to the Sheep Creek adit, which tunnels through greenschist under Mount Roberts to the shaft in the Silver Bow Basin. Since 1989 Echo Bay Mining Company has been working to reopen the Alaska Juneau Mine and plans to use the Sheep Valley to dump the 22,500 tons of waste rock that would be generated daily from this low-grade ore deposit.

The end of the Thane Road marks the start of Point Bishop Trail and foot access to Taku Inlet and the Taku River. This was a former trade route to the interior. North of Bishop Point is a transition in the degree of metamorphism in the rocks from greenschist to higher temperature and pressure amphibolite facies. Garnets appear in the amphibolite facies rocks, although most are microscopic.

The Taku River

Juneau is the only state capital in the United States that is inaccessible by road. Southeast of Juneau, the Taku River valley cuts through the coastal mountains, making a corridor to the interior. Juneau road builders have long considered this a potential highway link to the road systems of British Columbia. Unfortunately, the uncertainty about the behavior of Taku Glacier makes such a construction project very risky.

Taku Glacier is the largest glacier draining the Juneau Icefield and the only one that is now advancing. The thirty-mile-long glacier is advancing across Taku Inlet at a rate of a few hundred feet each year. If this continues over the next thirty to forty years, the glacier will cross Taku Inlet and form a very large ice-dammed lake. Old shorelines high up on both sides of Taku Inlet show that the glacier formed such a lake about 250 years ago when it advanced across the river mouth.

Old photographs of Juneau show icebergs in Gastineau Channel. They calved off Taku Glacier and floated down Taku Inlet into Stephen's Passage. Strong southeasterly winds created by high pressure areas moving out of the interior and down the Taku River valley may have helped to move the bergs northwest into Gastineau Channel. As the glacier advanced, it produced a large volume of sediment that began to fill the inlet. Since the 1950s, the end of Taku Glacier has rested on a moraine pushed up by its advance over its own sediments. This perch for the end of the glacier decreases the size of calved icebergs and reduces ice loss, a factor that contributes to the glacier's advance across Taku Inlet.

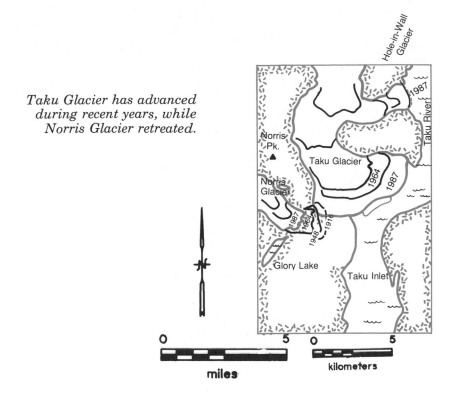

Taku Glacier has advanced during recent years, while Norris Glacier retreated.

Gastineau Channel and Douglas Island

This narrow passageway of Gastineau Channel was carved by ice that covered the Juneau area probably during much of the past 2 million years. The ice was hundreds of feet thick and it plucked at and removed the weakened bedrock along the Gastineau fault forming Gastineau Channel. Toward the end of the last major glaciation, 13,000 to 9,000 years ago, a thick layer of poorly sorted sand, gravel, and boulders was dumped from berg and shelf ice, and scrambled into the soft, fine-grained silty muds of Gastineau Channel. The resulting coarse- and fine-grained sediment, called a diamicton, now forms the surface material covering much of the Juneau-Douglas area. Beach deposits formed at elevations as high as 690 feet above modern sea level. As the load of ice melted, the land began to rise and sea level reached its modern elevation about 9,000 years ago.

Along the northeastern side of Gastineau Channel, Lemon Creek and the Mendenhall River are depositing their sediments to form deltas. These deposits will eventually fill the northwestern end of

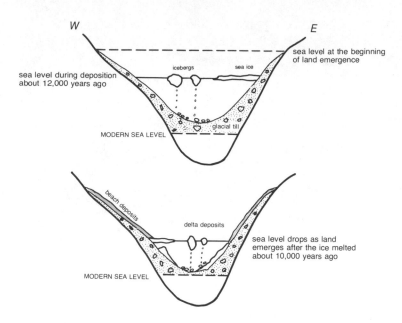

W   E

sea level at the beginning of land emergence

sea level during deposition about 12,000 years ago

icebergs   sea ice

glacial till

MODERN SEA LEVEL

beach deposits

delta deposits

sea level drops as land emerges after the ice melted about 10,000 years ago

MODERN SEA LEVEL

The complex dance of sea and land levels that helped make Gastineau Channel what it is.

Gastineau Channel to form a causeway joining Douglas Island to the mainland. At low tide the deltaic sediments form treacherous sand bars that force large cruise ships to go around Douglas Island in order to travel northward from Juneau to Auke Bay. North of Juneau, Egan Drive and the Glacier Highway provide access to some interesting geological features.

Douglas and the adjacent mining townsite of Treadwell were one of the most populous mining areas in Alaska in 1901 when a population of more than 1,000 people depended on the gold mines. Placer gold was first found on the beach in 1880, and in 1881 Pierre "French Pete" Errusard sold the first lode claim, the "Parris", to John Treadwell. The Treadwell Complex grew into a 240 and 300-stamp mill at the Treadwell Mine, as well as the "700 (foot) Mine", the Alaska Mexican, and the Ready Bullion Mines and its mills. The 300 (stamp) Mill was the largest in the world in its time. The old Glory Hole or open pit and many of the old mine buildings and mine sites can be seen by following the trail at the end of St. Anne's Road in Douglas.

To run the mills the mining company needed water to generate power. Between 1882 and 1890 an 18 mile long series of ditches was built along the 950 foot elevation beginning at Fish Creek in the

Eaglecrest Ski area on the northern end of Douglas Island. The ditch gathered water from Fish, Eagle, Kowee, Lawson, and Paris creeks, wrapping itself around the north and eastern side of the island to emerge at the 552 foot elevation above Treadwell. Two additional branches were brought in from Bullion and Ready Bullion creeks to the south of Treadwell. Water carried in the ditches supplied water wheels and steam plants to run the stamp mills that crushed the ore. A hike along the ditches today provides good exposures of the Douglas Island volcanic rocks near Fish Creek and the more metamorphosed greenschist along the eastern side of Douglas Island. Excellent exposures of slate are in West Juneau on the hillside above the Douglas Bridge.

West and North of Juneau

Juneau-Douglas High School on Egan Drive, 1.5 miles northwest of the downtown ferry terminal, is adjacent to the path of a very large avalanche chute at the base of Mt. Juneau. Near Gold Creek, along the mountain's southwest-facing slope, north of town above Wickersham Avenue, and along the southwest-facing slopes of Mt. Roberts above Thane Road are additional avalanche chutes. A major slide in 1962 brought hundreds of tons of snow down the Behrend's Avenue chute and reached Gastineau Channel, dusting the boats in Harris Harbor. Nearly two decades later during the winter of 1985, a heavy snowfall accumulation on Mt. Juneau ridges was released suddenly.

1985 avalanche off Mt. Juneau onto Wickersham Avenue, Juneau.
—Rod Flynn photo

One avalanche trapped a woman in her home on Wickersham Avenue, another produced an instant snowboulder wall adjacent to a house on Judy Lane, and a third narrowly missed a mining museum caretaker as she walked down Basin Road.

At Eaglecrest Ski area in the Fish Creek Valley on Douglas Island, blasting techniques are used to induce controlled avalanches that might otherwise engulf unwary skiiers. The City and Borough of Juneau instructed the planning commission to deny building permits in avalanche zones. Unfortunately it allowed construction in already platted areas.

At mile 3.8 on the northwest side of the Egan Highway is Salmon Creek. The Wagner Mine staked in 1898 on Salmon Creek never produced much gold. Two large pelton wheels, all that remain of the old two-stamp mill, are still visible in front of the waterfall. At the head of the valley is Salmon Creek Dam and reservoir. Built in 1913-1914 to supply power to the A-J Mill, the reservoir is now used as a source of drinking water for the City and Borough of Juneau. On the northern side of Salmon Creek Valley is Blackerby Ridge that separates Salmon Creek from Lemon Creek. It provides access to Camp Seventeen on the Juneau Icefield used by Professor Maynard Miller and his students for icefield studies.

At mile 5.9 Lemon Creek enters Gastineau Channel. Here bald eagles are commonly seen along the creek and around the piles of sand and gravel mined from two pits near the creek mouth. The creek was originally mined for placer gold by John Lemon in 1897. By 1983, 252,000 tons of sand and gravel were mined from these pits. In Alaska, landscapes covered by bogs and permafrost make sand and gravel deposits more valuable than gold because nothing can be built without first creating a building pad of these materials.

Lemon Creek Glacier at the head of this valley is one of the better studied glaciers in the Juneau Icefield. The glacier has retreated more than one and one-half miles over the last 230 years.

Juneau Icefield

Stretching along the crest of the Coast Range from the Taku River to just east of Skagway at the south end of Lake Atlin, the Juneau Icefield is one of the most studied icefields in the world. It covers an area of more than 1,200 square miles in Alaska and extends into British Columbia. Annual snowfall of more than 100 feet on the Juneau Icefield supplies more than 30 valley glaciers that descend from the main ice mass to near sea level. At depths of 200 feet or more the ice crystals fuse together under the pressure of the overlying ice and flow plastically down valley. Ice is frequently lost at the glacier's terminus through calving or breakage. During years of low snowfall the supply of ice is not as great in the source area and tributary glaciers are not resupplied. With the exception of Taku Glacier, all glaciers draining the Juneau Icefield are retreating.

Mendenhall Glacier

Mendenhall Loop Road leads to the Mendenhall Glacier and visitor center. On the right, forming the southeastern side of Mendenhall Valley, is Thunder Mountain, a spur of Heintzleman Ridge. From the top of this ridge are excellent views of the glacier and Mendenhall River system. Siltstone and mudstones in Heintzleman Ridge have been heated to produce metamorphic minerals such as biotite and garnet.

The modern Mendenhall Glacier is a relic of the Little Ice Age that began about 3,000 years ago. Since about 1750 the Mendenhall Glacier has been melting back up the valley. It is still 12 miles long, 1.5 miles wide, and more than 100 feet high at its terminus.

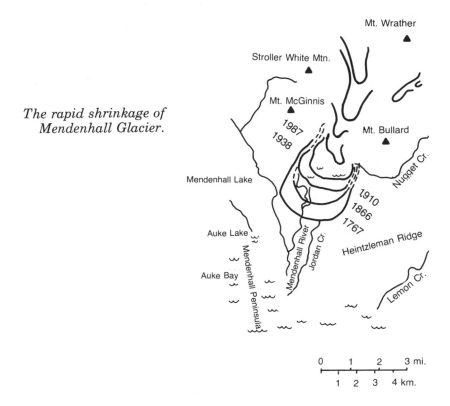

The rapid shrinkage of Mendenhall Glacier.

Saltmarsh vegetation has stabilized the braided outwash fan on the tidal flats below Mendenhall Glacier. The Mendenhall River is developing a meandering stream channel. —Rod Flynn photo

The Nugget Creek power plant was built between 1912 and 1914 at what was then the terminus of Mendenhall Glacier to produce hydroelectric power for the Treadwell Mines in Douglas. The powerhouse was connected by a 6902-foot flume and pipeline system to Nugget Creek on the east side of Mendenhall Glacier and produced 2300 kilowatts under peak conditions.

Mendenhall Glacier is one of the best sources of single crystal ice samples used for laboratory research. It ends in Mendenhall Lake, which is 200 feet deep at its maximum depth. Most of the lake did not exist before the 1930s because ice filled the basin. The Visitor Center site was still covered by ice in 1940. The current rate of retreat is less than 100 feet per year. The ice still flows forward at a rate less than 5 feet per day.

Before 1750, Mendenhall Glacier reached 2.5 miles farther down the valley. Thirteen different recessional moraines have been recognized between this end moraine and the Visitor Center. Each records a temporary pause in the retreat of the glacier since that time. The moraine trail provides excellent views of newly deglaciated surfaces and the slow process of revegetation. Lichens first break down the rocks to produce a fine soil, pioneering plants then appear such as willow, fireweed, and alder. Forests of Sitka spruce and western hemlock grow about 75 years after the ice melts off the land.

Out The Road

Ten miles northwest of Juneau, the Brotherhood Bridge crosses the Mendenhall River. During high tides, salt water moves up the Mendenhall River as far north as its confluence with Montana Creek.

The layered sediments exposed in the cutbank on the west side of the river near the bridge show the changing depositional environments from estuarine, to fluvial, to wet meadow. This transition took place as sea level rose after the end of the Ice Age, and the land rebounded upward in response to the removal of thousands of feet of ice. Large gravel bars in the Mendenhall River have provided construction material for many Juneau projects. Peat covered by glacial and marine deposits along Montana Creek is older than 39,000 years. It contains fossil pollen that records a plant community dominated by sage and sedge, with small amounts of pine, fir, hemlock, spruce, and alder. The climate was cooler at that time, and these peats may have formed when glaciers receded out of the inner coastal areas of southeast Alaska.

Auke Lake, once a salt chuck connected to Auke Bay 10,000 years ago, now provides a spectacular setting for the University of Alaska. To the northeast, the rocks are greenschist grade. To the southwest, excellent outcrops along Fritz Cove Road are mudstones. Smuggler's Cove, at the end of Fritz Cover Road near Spuhn Island, exposes steeply dipping slates interbedded with conglomerate that also appears farther north near Eagle Beach. Near the Auke Bay Ferry Terminal, a large bedrock outcrop above the road is composed of graywacke and volcanic tuff. To the northeast behind Auke Bay and Auke Nu Mountain, the Gastineau Channel fault follows the trend of the Montana Creek and Windfall Lake valleys northward into Berner's Bay.

At about Mile 23, the turn-off for the Shrine of St. Therese leads to a small rocky crag attached to the mainland by a narrow causeway. The Shrine stands on volcanic tuff and breccia formed more than 60 million years ago as explosive fragments from a volcano that later consolidated and hardened into rocks. The Chilkat Range lies across Lynn Canal to the northwest. Between 1897 and 1922, the John Peterson family operated a small gold mine and mill about a half mile east of Peterson Lake at the headwaters of Peterson Creek. Follow the trail about 4 miles back into the hills from Mile 24 on the Glacier Highway.

At the mouth of Herbert and Eagle rivers, a salt mash was formed on sediment washed down by the rivers. Two hundred years ago at the end of the Little Ice Age, relative sea level was about 20 feet higher than present. The salt marsh at that time formed 35 feet above present sea level. The salt marsh of today is forming 15 feet above sea level. The present site of Boy Scout Camp and the surrounding forest is on a 20 foot terrace that marks the site of the Little Ice Age intertidal flat that tree ring counts date to 1765.

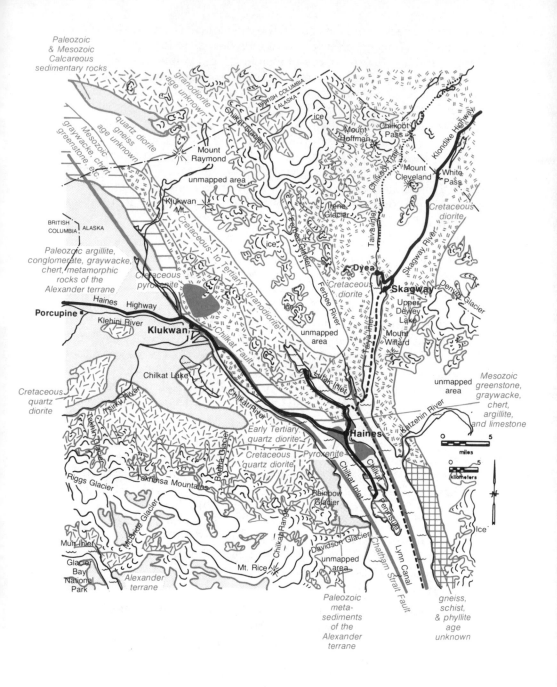

Paleozoic & Mesozoic Calcareous sedimentary rocks

quartz diorite age unknown

gneiss age unknown

Mesozoic graywacke, greenstone, etc.

'granodiorite' age unknown

Chilkat Glacier

ice

Mount Raymond

Mount Hoffman

Chilkoot Pass

Klondike Highway

unmapped area

Mount Cleveland

White Pass

Klukwan Mt.

Irene Glacier

Cretaceous diorite

BRITISH COLUMBIA ALASKA

BRITISH COLUMBIA ALASKA

Cretaceous to Tertiary

ice

Ferebee Glacier

Skagway River

Paleozoic argillite, conglomerate, graywacke, chert, metamorphic rocks of the Alexander terrane

Cretaceous pyroxenite

Dyea

Cretaceous diorite

Skagway

Denver Glacier

Haines   Highway

'granodiorite'

ice

Upper Dewey Lake

Porcupine

Klehini River

Mount Willard

Mesozoic greenstone, graywacke, chert, argillite, and limestone

Klukwan

Chilkat Fault

unmapped area

unmapped area

Cretaceous quartz diorite

Chilkat Lake

Chilkat River

Chilkat Inlet

Katzehin River

Takhin River

Tsirku River

Early Tertiary quartz diorite

Pyroxenite

Haines

0   5
miles

Cretaceous quartz diorite

Bertha Glacier

Takhinsa Mountains

0   5
kilometers

Riggs Glacier

McBride Glacier

Chilkat Range

Chilkat Inlet

Rainbow Glacier

Chilkat Peninsula

Chatham Strait Fault

Lynn Canal

Ice

Muir Inlet

Glacier Bay National Park

Alexander terrane

Mt. Rice

Davidson Glacier

unmapped area

Paleozoic meta-sediments of the Alexander terrane

gneiss, schist, & phyllite age unknown

HAINES—SKAGWAY REGION
Klondike and Haines Highways

78

Juneau (Auke Bay)—Haines
Sailing time: about 5 hours
80 air miles

The prominent outcrop next to the road just east of the ferry terminal exposes volcanic ash, sandstone, and fine-grained sediments now slightly metamorphosed. Auke Nu Mountain, the prominent knob north of the ferry terminal, is also made of slightly metamorphosed sediments. Rocks north and east of Montana Creek along the Gastineau Channel fault, are more thoroughly metamorphosed. Step-like ridges that parallel the mountainside were etched by Ice Age glaciers and later covered by rain forest. Auke Bay was deeply scoured by glacial ice to make an excellent site for a ferry terminal.

The ferry sails west out of Auke Bay, passing Coghlan Island with its rocky outcrops of metamorphosed silts and clays. To the west Mansfield Peninsula on the northern tip of Admiralty Island, is marked by Mt. Robert Barron at 3450 feet. It is snow-covered much of the year.

About 10 minutes from the terminal the ferry passes the northeast end of Spuhn Island, composed of metamorphosed sandstone, and siltstone. The ferry next swings north parallel to Portland Island and follows the east side of Shelter Island, passing the neighborhoods of Lena Point and Tee Harbor on the mainland. The rocks here are metamorphosed sandstone, graywacke, conglomerate, and some basalt.

After about an hour the ferry nears the north end of Shelter Island where prominent east-west-trending scours on the upper peaks indicated the work of glacial ice. Deep notches on adjacent Lincoln Island record the passage of the 2,000 foot thick ice blanket that once covered this part of the southeastern Alaska panhandle.

Herbert and Eagle River Systems

On the mainland, the Herbert and Eagle river systems have a combined delta at Eagle Beach. As the delta builds outward and upward, it provides a foundation for salt marsh plants. Two hundred years ago at the end of the Little Ice Age, sea level and the salt marsh were about 20 feet higher than present. The Eagle River mine property was worked between 1903 and 1923 and produced about 75,000 tons of gold ore.

Herbert and Eagle Glaciers

Both the Herbert and Eagle rivers begin at melting glaciers. Herbert Glacier advanced over a forest probably during the Little Ice Age between about 1500 and 250 years ago. The river is now exhuming the stumps of this fossil forest in a prominent river cutbank about a mile up the Herbert Glacier Trail. Since 1766 the glacier has retreated an average of 60 feet per year, with a maximum average rate of retreat of 190 feet per year between 1928 and 1948. The Herbert Glacier trail crosses a series of morainal ridges left by the receding glacier, the tallest of which is about 20 feet high. Pioneering plants appear 3 to 7 years after ice has formed moraines. Lupine and dwarf fireweed are among the first to appear, along with seedlings of alder, willow, cottonwood, and Sitka spruce. Alder thickets predominate for the next 50 years until spruce outgrow the alders and begin conifer forest. Herbert Glacier began to recede from its most recent advance in the Little Ice Age about 1765 A.D., judging from growth ring counts of conifers growing in this moraine.

Berner's Bay

Hardrock gold was first discovered in the Berner's Bay region in 1886 or 1887 by prospectors who followed ore-bearing stream cobbles upstream along Sherman Creek to outcrops in the upper portion of the drainage basin. This is the northern end of the Juneau gold belt, which extends southeastward through Juneau and about 100 miles southeast to Windham Bay. By the end of 1909, approximately $1,100,000 in gold ore had been recovered from five stamp mills in the Berner Bay region. The Comet and Jualin mines supplied most of the ore. The settlement of Comet on Lynn Canal was once a regular stop for local steamers operating between Juneau and Skagway.

Chilkat Range

The Chilkat Range on the west side of the Chatham Strait fault belongs to the Alexander terrane, late Cambrian to Triassic metamorphosed sediments described in the Glacier Bay segment. These rugged peaks are snow-covered much of the year and are still being mapped by Alaskan geologists with the help of helicopters.

Rocks on the eastern side of Lynn Canal are more severely metamorphosed than those in the Chilkat Range. This belt of metamorphosed rocks may once have been part of the Alexander terrane. Now it forms an envelope around the granitic rocks of the Coast Range that outcrop in the Skagway area. The prominent sharp peaks along the ridgeline were once nunataks that stood above the Pleistocene Juneau Ice Field. The word nunatak originated with the Eskimo people and means "lonely peak."

George Vancouver named Lynn Canal after his birthplace in England. The absence of glacial fjords in England must be the reason Vancouver failed to recognize that his so-called canal was not man-made, but carved by glacial ice. When the glacier ice melted, the sea level rose, this former river valley was drowned by sea water to form a 1,000-feet-deep, 3-to-6-mile-wide, 60-mile-long fjord. The Chatham Strait fault fractured the rocks before Tertiary time, creating a zone of weakness along which the glacial ice could more easily erode. The Chatham Strait fault and Gastineau Channel fault intersect just south of Chilkat Peninsula, under the waters of Lynn Canal.

Chilkat Peninsula

Davidson Glacier comes into view near the south end of the Chilkat Peninsula. This finger of land borders the Chilkat River and Chilkat Inlet on the east and indicates we will soon be in Haines. This country is the domain of the Chilkat and Chilkoot clans of the Tlingit people. For many years they controlled trade with the interior and were pretty tough customers for the early European explorers trying to get a foothold along the Alaska coast.

Note the steep walls of the Lynn Canal fjord at the south end of Chilkat Peninsula. Many Alaskans dream of a road link between Juneau and Haines along the western, Chilkat side of Lynn Canal with a bridge near Berner's Bay. Active glaciers, strong winter storms, and declining oil revenues have delayed the bulldozers for the time being.

Haines—Skagway
Sailing time: about 1 hour
10 air miles

The last or northernmost stop for boat traffic in Lynn Canal is Skagway. Proceeding up Lynn Canal into Taiya Inlet from Haines, the rock walls on either side of the fjord are very steep with waterfalls. "Skagway", home of the north wind in the Tlingit language, is an appropriate name for this town. Klondike gold seekers discovered this north wind in 1898 when Skagway's population swelled to 20,000 people. The Great Depression of 1897 in the lower 48 states fueled the gold fever that brought hordes of people north to struggle over the Chilkoot Trail into the Yukon River watershed and on downriver to the goldfields of Dawson.

SOUTHEAST ALASKA
AIR AND BOAT ROUTES

Glacier Bay National Park

From Juneau regularly scheduled and charter flights are available to Gustavus Airport and Glacier Bay, across the Chilkat Range approximately 100 miles to the northwest. Water access is either from the Pacific Ocean or by way of the Inland Passage through Icy Strait. Cruise ships make regular visits into Glacier Bay during the summer.

By Plane from Juneau

From Juneau Airport the flight line to Glacier Bay passes over the Chilkat Range. These mountains are Paleozoic limestone, marble and metamorphic marine sedimentary rocks of the Alexander terrane. As the plane nears Gustavus Airport a large foreland appears just east of the Park boundary. This glacial outwash plain was produced by Little Ice Age glaciers between about 1400 and 1750 A.D., when ice completely filled Glacier Bay. Today tall lodgepole pines thrive in Gustavus and line the roadway into the park. Near the park boundary the road passes over glacial moraines and the pines are replaced by 150 year-old forests of Sitka spruce and hemlock, with a mossy understory. Spruce trees that began growing here 100 years ago are now 120 feet tall. They replaced thickets of alder that colonized the area soon after deglaciation and built up the soil by fixing nitrogen. Hemlocks will eventually replace the spruce as forest succession continues.

Park headquarters is at Bartlett Cove near the entrance to Glacier Bay. In 1794 when George Vancouver sailed by the entrance to Glacier Bay, Bartlett Cove was just emerging from glacial ice. Vancouver's maps show that the present lowland areas on either side of the entrance to Glacier Bay have risen or "rebounded" between 18 and 21 feet since the ice melted about 190 years ago.

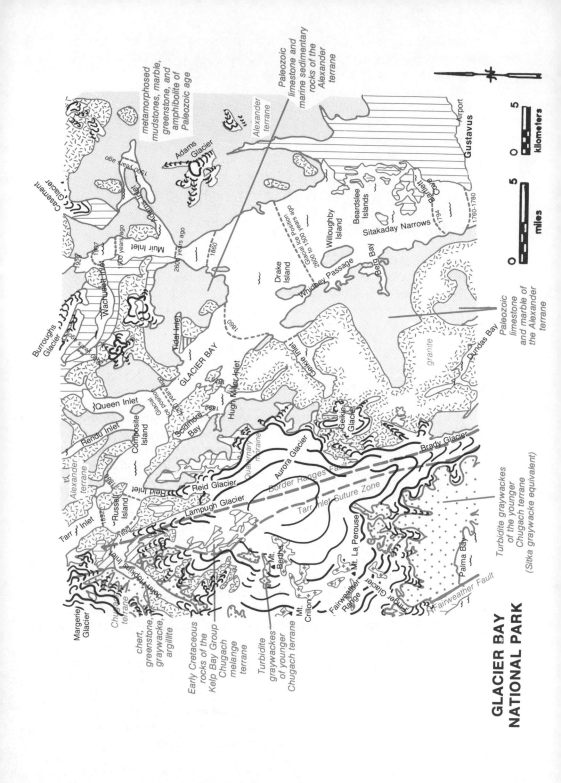

GLACIER BAY NATIONAL PARK

Paleozoic limestone and marine sedimentary rocks of the Alexander terrane

metamorphosed mudstones, marble, greenstone, and amphibolite of Paleozoic age

Alexander terrane

Airport

Gustavus

kilometers

miles

Casement Glacier

1300 years ago

Adams Glacier

Adams Inlet

Beardslee Islands

Bartlett Cove

Bartlett Inlet

1794

1760-1780

Muir Inlet

500 years ago

1907

1929

Sitakaday Narrows

Willoughby Island

Whidbey Passage

260 years ago

Glacial Ice Position 2600 to 3500 years ago

500 years ago

1860

Wachusett Inlet

Burroughs Glacier

Drake Island

Berg Bay

250 years ago

1860

Tidal Inlet

GLACIER BAY

Paleozoic limestone and marble of the Alexander terrane

granite

Dundas Bay

Queen Inlet

Ice position 1750 years ago

Composite Island

Rendu Inlet

Alexander terrane

Sebree Island

Sandy Cove

Hugh Miller Inlet

Geikie Inlet

Scidmore Bay

Reid Glacier

1892

Reid Inlet

1892

Russell Island

1892

Tarr Inlet

Aurora Glacier

Geikie Glacier

Quaternary moraine

Brady Glacier

Lampugh Glacier

Border Ranges Fault

Tarr Inlet Suture Zone

John Hopkins Inlet

Chugach terrane

Margerie Glacier

chert, greenstone, graywacke, argillite

Early Cretaceous rocks of the Kelp Bay Group Chugach melange terrane

Turbidite graywackes of younger Chugach terrane

Mt. Bertha

Mt. Crillon

Fairweather Range

Mt. La Perouse

Mt. Fairweather

Finger Glacier

Palma Bay

Fairweather Fault

Turbidite graywackes of the younger Chugach terrane (Sitka graywacke equivalent)

Glacier Bay National Monument was established in 1924 to preserve an area of Alaska that is remarkable both for its scenery and its glacial history. In 1980, Congress renamed the area Glacier Bay National Park and Preserve, a total area of about 3.2 million acres. It is now closed to mining. The park is bordered by the Chilkat Range to the east, the Takhinsha Mountains on the north, the Alsek and St. Elias ranges to the northwest, and the Gulf of Alaska to the southwest.

The St. Elias Mountains include Mt. Logan, at 19,850 feet the tallest peak in Canada and the second tallest in North America. They can be seen from the visitor center porch on a clear day. The majestic Fairweather Range culminating in Mount Fairweather at 15,300 feet forms an impressive divide between the Gulf of Alaska and Glacier Bay. It is visible from the West Arm of Glacier Bay. From Park Headquarters, one can charter a small tour boat or bring one's own folding kayak to begin a journey into one of the world's most spectacular areas.

Glacial History

The Tlingit Indians once lived on the Beardslee Islands northwest of Bartlett Cove and Park Headquarters. A tribal story relates how they were forced to move south across Icy Strait to their present village of Hoonah when a glacier advanced over its alluvial plain toward the entrance of Glacier Bay. The missionary Hall Young, who accompanied John Muir from Wrangell into Glacier Bay by canoe in 1889 and 1880, was asked by one of the Hoonah subchiefs to pray for removal of the ice mountain that had overridden and spoiled his king salmon stream. Eight years later the glacier began to melt back, greatly strengthening Young's stature with the locals.

The Little Ice Age began about 1500 years ago and glaciers advanced into the bay until they filled it with ice. By the time George Vancouver sailed through Icy Strait in 1794, a bay approximately 5 to 7 miles long had opened behind the melting ice. The Russian Tebankov, who produced the first "Atlas of the Northwest Coast of America" and was governor of the Russian American colonies between 1845-1850, named Icy Strait in 1852 after the icebergs.

The glaciers in Glacier Bay have been retreating steadily since the Little Ice Age. In just over 200 years they have melted back more than 65 miles, leaving the series of spectacular fjords that make up Glacier Bay. Various glacial deposits formed during this time; moraines, terraces, channels, and ridges form a barren chaotic landscape that provides clues to unraveling the glacial history of the area.

By Boat into the
West Arm of Glacier Bay

From Bartlett Cove the boat travels across the entrance to Glacier Bay and follows the western side of the bay. The gray limestone of Willoughby Island was polished by Little Ice Age glaciers. The rock formed during Devonian or Silurian time, about 400 million years ago, and is part of the Alexander terrane.

To the northwest along the western shoreline the boat passes a series of fossil tree stumps that record an ancient forest overrun by glacial ice. Watch for bears feeding along the shoreline in the early evening.

Past Marble Island, Tlingit Point comes into view on the eastern shoreline. In about 1860, Tlingit Point headland split the retreating glacial terminus into the Grand Pacific Glacier, which retreated to form the West Arm of the bay, and Muir Glacier, which retreated northward to form Muir Inlet. Before then, an ice wall fifteen miles long and about 250 feet tall formed the end of this huge glacier. After his early canoeing adventures, John Muir returned to Glacier Bay by steamer in 1890 and camped at Tlingit Point, then at the end of Muir Glacier.

From Tlingit Point you can see Casement Glacier in Muir Inlet to the northeast. The Fairweather Range lies off to the west. The prominent peaks of Mt. LaPerouse, Mt. Crillon, and Mt. Bertha are layered gabbro that was intruded into the Chugach terrane. Important nickel and copper deposits have been found in the gabbro.

Fairweather Range

In the western portion of Glacier Bay National Park the rocks of the Fairweather Range consist of phyllite, schist, not-quite-schist, and gneiss, all part of the Chugach terrane. These rocks correlate with Cretaceous muddy sandstones that extend west along the south Alaska coast all the way to the Shumagin Islands.

Fairweather Transform Fault

The Fairweather fault separates the terranes of the North American Plate from the Pacific Plate and the tiny Yakutat block, which is now docking onto North America. The Yakutat block contains Chugach terrane rocks that are overlain by Tertiary sedimentary and volcanic rocks. Northeast of the monument the Fairweather transform fault connects with the eastern end of the Aleutian trench near Icy Cape.

Geike Inlet

Between about 8,000 and 2,500 years ago the glacier in Glacier Bay advanced to Geike Inlet, blocking Muir Inlet with an ice dam to create a glacial lake. The Van Horn formation exposed in Muir Inlet contains sediments deposited in that lake. Water in the West Arm is now more than 1200 feet deep. Between 1860 and 1880 the Grand Pacific Glacier retreated twenty miles up the West Arm from this area.

Earthquakes and Glaciers

An earthquake triggered a landslide in glacial sediments at Tidal Inlet, in West Arm. George Vancouver and his crew felt an earthquake in 1794, which was later blamed for the withdrawal of ice from Glacier Bay. Other scientists later thought that earthquakes were responsible for the sudden retreat of tidewater glaciers.

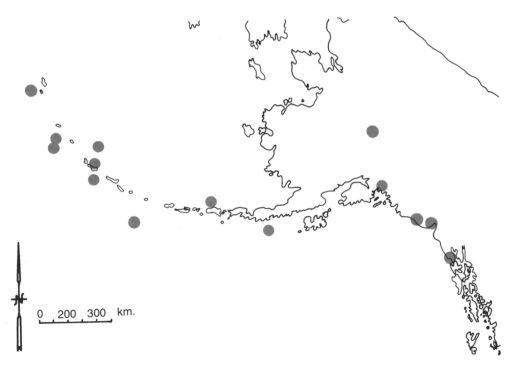

Epicenters of great earthquakes in Alaska from 1899 to 1964.

The idea went out of favor when geologists learned that earthquake waves do not penetrate ice very well, and thus cannot directly cause change in the mass balance of glaciers. Glaciologists discovered this when they tried to use man-made seismic waves to determine thickness of ice in places like Greenland and Antarctica. Now they use radar, which is able to penetrate the ice. After the 1964 Good Friday earthquake in Alaska, glaciologists looked at tidewater glaciers near the epicenter of the earthquake to see if there had been any effect. None was found.

Glacier Bay Intrusive Rocks and Mineral Deposits

The intrusive rocks of the monument include layered gabbros of unknown age and granite of mid-Tertiary age that form north-trending belts in the western third of the park; they parallel Tarr Inlet and the trend of the Fairweather transform fault. Mid-Cretaceous granitic rocks exist in both the north-trending belt east of the layered gabbro and Tertiary granite, and in a west-northwest-trending belt in the northeastern third of the park.

Known mineral deposits within Glacier Bay National Monument.

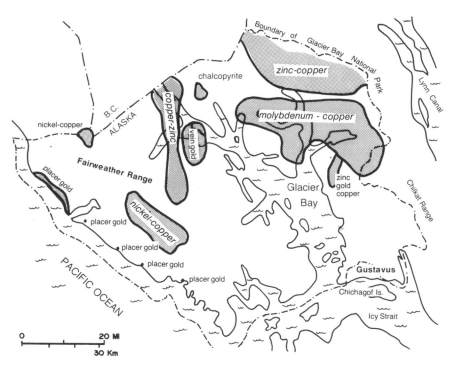

Within and adjacent to these intrusive rocks lies the mineral wealth of the park. Along the Pacific coast, beach sands contain placer gold derived from the Fairweather Range. Gold has been found in veins near Reid Inlet. The Crillon-Laperous area contains a magmatic nickel-copper deposit as does Mt. Fairweather near the British Columbia border. Margerie Glacier and Tarr Inlet contain porphyry copper deposits. Orange Point contains zinc and copper deposits. A massive chalcopyrite skarn deposit lies near Rendu Inlet.

Reid Inlet

Reid Inlet on the southwest side of the West Arm is a popular spot for dinner stops and overnights. Here Reid Glacier at the head of the inlet is retreating almost entirely through melting and evaporation. The glacier calves very few icebergs, and has a rounded profile. An actively calving glacier shows blue ice at its terminus. The white ice at the end of Reid Glacier suggests that it is not an iceberg producer.

Two streams at the edge of the glacier carry large quantities of sediments into the head of the inlet, as much as 12 feet per year. Core samples taken from the inlet contain mostly glacial rock flour or silt. Thin layers of sand 1 cm thick are near the deltas of the streams. These layered sediments look very much like annual varve layers produced in lakes. It is sometimes difficult for geologists to distinguish between lake sediments and glacially derived sediments deposited in sea water unless they contain fossils.

Tarr Inlet and Plate Tectonics

The boat now heads northwestward into Tarr Inlet. Rocks of the Chugach terrane are to the south, much older rocks of the Alexander terrane lie along the north side of the fjord. Glacier Bay National Park is at the northern end of the Alexander Archipelago. Plate tectonic interpretation suggests that the Precambrian through middle Paleozoic rocks of the Alexander terrane, comprising the eastern two-thirds of the park, were joined to a Paleozoic North American plate about 250 million years ago. The suture lies in interior British Columbia.

In Tarr Inlet, rocks of the Alexander terrane are bordered to the west by the Tarr Inlet suture zone, a narrow strip of low-grade metamorphic rocks like phyllite, slate, conglomerate, chert, greenstone, greenschist, and marble. The suture zone rocks are structurally complex, and between about 230 and 90 million years old.

They occur in discontinuous units that correlate with the Kelp Bay Group on Chichagof Island, a melange in the Chugach terrane. West of the suture zone are the homogenous muddy grey sandstones of the Chugach terrane.

The northeastern margin of the Tarr Inlet suture zone is the Border Ranges fault, which continues northwest across southern Alaska and elsewhere separates the Wrangellia and Chugach terranes. Rocks of the Wrangellia terrane have not been identified in Glacier Bay, but they do exist south of Lisianski Inlet. Thus early Cretaceous rocks of the Chugach terrane are juxtaposed against much older rocks of the Alexander terrane. Jurassic and Cretaceous granite intrude the suture zone.

Grand Pacific and Margerie Glaciers

By 1992 the Grand Pacific and Margerie Glaciers had rejoined after a 70 year separation. The last joining of these two glaciers was in 1912 when Grand Pacific Glacier began to retreat into Canada. The Canadians entertained the notion of building a seaport at the head of the inlet but by 1966 the glacier had readvanced over the international boundary back into the U.S. In the spring the Kittiwake, a type of gull, raise their young in a rookery in the rocky cliffs near these two glaciers. The clever birds wait for ice to calve off the glacier front and then feed on the shrimp and krill churned up in the turbulent water.

Very poorly sorted glacial sediments are produced in Tarr Inlet as icebergs roll over, dumping their gravel onto the fjord floor, or as stream sediments quickly mix with fine-grained silt and clay that settles slowly out of the water column.

John Hopkins Glacier

Striped icebergs made up of layered ice and glacial sediments originate in John Hopkins Glacier along the northwest side of the Tarr Inlet. Harbor seal females and their pups live on them. In May 1984, at Lampugh Glacier on the southwest side of John Hopkins Inlet, a river was seen pouring out of the center of the ice front well above water line much like a giant spigot. This is most unusual.

Rendu Inlet,
Glacier Bay,
July 31, 1906.
—U.S. Geological Survey
photo by C.W. Wright

Along the edge of the Tarr Inlet suture zone are the LeRoy and Rainbow gold mines; a copper-zinc prospect lies across Johns Hopkins Inlet to the north. Rocks next to Lampugh glacier are rainbow-colored, showing signs of alteration by hot ground water at depth. Here also plants have begun their work of recolonizing the recently deglaciated landscape. First lichens break down fresh rock to create soil for the small herbaceous plants, dryas, willow, and dwarf fireweed, which follow. Alder will come in later.

A notch north of Rendu Inlet offers a view of Carroll Glacier, which originates in British Columbia. Twenty years ago Plateau Glacier exited from Carroll Glacier and filled Queen Inlet, which is now ice-free. As you pass Queen Inlet you can see that Carroll Glacier has retreated and is now grounded.

Watch for tufted puffins and diving cormorants. The cormorants will dive as deep as 15 feet for food. Humpback whales and killer whales are common visitors to the park in the summer months. Hydrophones were strung up around Glacier Bay to measure ship noises and their effects on whales.

1906 photo by
Carroll Glacier,
Queen Inlet,
Glacier Bay.
—U.S. Geological Survey
photo by C.W. Wright.

91

Muir Inlet

Glaciers in Muir Inlet began to retreat before 8,000 years ago when temperatures were probably warmer than present. Geologists have studied the outwash gravels deposited by the melting glaciers and found tree stumps and forest sediments dating back about 8,000 to 3,000 years. Lake sediments began to accumulate on top of the gravels between about 2,900 and 2,300 years ago. A glacier advanced down the west arm of Glacier Bay and blocked Muir Inlet to impound the lake. Between 1860 and 1880, after the Little Ice Age, Muir Glacier retreated only two miles up its narrow inlet compared to the twenty miles of retreat by Grand Pacific Glacier in the West Arm during the same time period.

1954 aerial view of Lituya Bay flanking the Fairweather Range. The Fairweather Fault trace is marked by the steep cliffs at the head of the bay. The steep slopes bordering the bay were recently glaciated and have been scoured repeatedly by giant earthquake–generated waves. The morainal ridge at the mouth of the bay has been breached by the sea. —U.S. Geological Survey photo by D.J. Miller

Glacier Bay—Yakutat— Cordova—Valdez

Some of the world's most spectacular geology passes beneath the airplane window between Juneau and Anchorage, along the south central coastline of Alaska. An Alaska Airline's "milk run" that stops in the towns of Yakutat and Cordova is an excellent way to see this country—when the weather is clear.

Lituya Bay

Between Glacier Bay and Yakutat is the unusual, 7 mile long, T-shaped Lituya Bay. This tidal inlet was scoured by ice as much as 720 feet below present sea level. At that time sea level was much lower. Melting Ice Age glaciers later raised sea level to drown the valleys formerly occupied by ice. The narrow entrance to Lituya Bay is only 33 feet deep. The T-shaped head of the bay lies along the trench of the Fairweather fault at the foot of the Fairweather Range and Saint Elias Mountains.

In 1958 a major earthquake centered near Lituya Bay loosened about 40 million cubic yards of rock in a massive rockslide. The rocks roared ino Gilbert Inlet at the northwestern head of Lituya Bay from an altitude of about 3,000 feet on the steep northwest wall. The belly flopping rockslide raised a wave that surged over the opposite wall of

1985 aerial view of Lituya Glacier and wave damage west of Gilbert Inlet in Lituya Bay induced by earthquake-generated rockslide into bay. –U.S. Geological Survey photo by D. J. Miller

the inlet to a height of 1,720 feet. A huge wave moved down the bay to its mouth at between 97 and 130 miles per hour. Two of three fishing boats in the outer part of the bay sank, and two people were killed. The water that sloshed against the steep-sided walls of the upper bay destroyed 4 square miles of forest.

Near Crillon Lake just southeast of Lituya Bay geologists have measured horizontal offset to the right along the Fairweather fault of 3.6 miles and vertical offset of 3 feet. Radiocarbon dates on fossil wood from glacial moraines near Crillon Lake and at Finger Glacier indicate that stream drainages offset by the fault are younger than about 1300 years old. Information from streams offset by the fault tell geologists that the average rate of movement along this part of the Fairweather fault for the past thousand years has been at least three inches per year, about average for plate movement. All over the earth's surface plates are moving at approximately this speed, about as fast as your fingernails grow.

Recent Activity on the Fairweather Fault

The northwest-trending Fairweather fault has been sliding the Pacific plate past the North American plate for the past 35 million or so years. The latest pulse of movement occurred quite recently. During the Lituya Bay earthquake of July 10, 1958, the entire 170 mile onshore length of the fault probably moved.

The onshore estimated rate of movement along the Fairweather fault is about the same as that between the Pacific and North American plates in the Gulf of Alaska, as calculated from deep sea fossil magnetic information. The Fairweather fault is an active transform fault along which most of the relative motion between the Pacific and North American plates is taking place. Large valleys that lie across the Fairweather fault and are older than 120,000 years are each offset to the right an average of 3.3 miles. These relocated topographic features suggest that the modern high rate of movement along the fault could not have started more than 100,000 years ago. Before that time the relative motion between the North American and Pacific plates focused along several offshore faults.

The Yakutat Block

The Yakutat block is a terrane 360 miles long and 120 miles wide that is now docking with southern Alaska. Rocks of the Yakutat block are a basement of late Mesozoic sedimentary formations in the eastern third of the block and early Tertiary oceanic crust in its western two thirds. Tertiary sediments overlie the basement rocks. Structural and earthquake seismic information indicate that the Yakutat block is moving along at the tectonically steady rate of about two and one half inches per year, probably for most of the last 100,000 years. Thick piles of Miocene and younger sediments fill basins in the continental shelf along the Gulf of Alaska. These sediments were eroded from the rising Saint Elias and Chugach Mountains along the northern margin of the Yakutat block.

The most likely starting point for the Yakutat block is the edge of the continent in southeastern Alaska and northern British Columbia. Plant and animal fossils in the early Tertiary layered sediments of the Yakutat block record a northward displacement from the latitude at which they originally lived.

Putting together all that is currently known, it looks like the Yakutat block was sliced off the continental margin southeast of Chatham Strait about 25 million years ago, and moved about 330 miles northwest along the Queen Charlotte-Fairweather transform fault system. At the same time, 540 miles of right-lateral displacement occurred along the Transistion fault system between the southern edge of the Yakutat block and the Pacific plate.

Yakutat

Yakutat stands at the mouth of Yakutat Bay, at the edge of the Yakutat forelands, and at the foot of the St. Elias Mountains. On June 27, 1984 a magnitude 6.2 earthquake, centered just west of Yakutat, raised the floor of Yakutat Bay. The cruise ship Cunard Princess stopped sailing into the bay when the captain noticed the water in some places was 20 feet shallower than his navigation charts indicated.

The Alsek Glacier fronts the eastern bank of the Alsek River upstream from its mouth at Dry Bay along the Gulf of Alaska.
—National Park Service photo

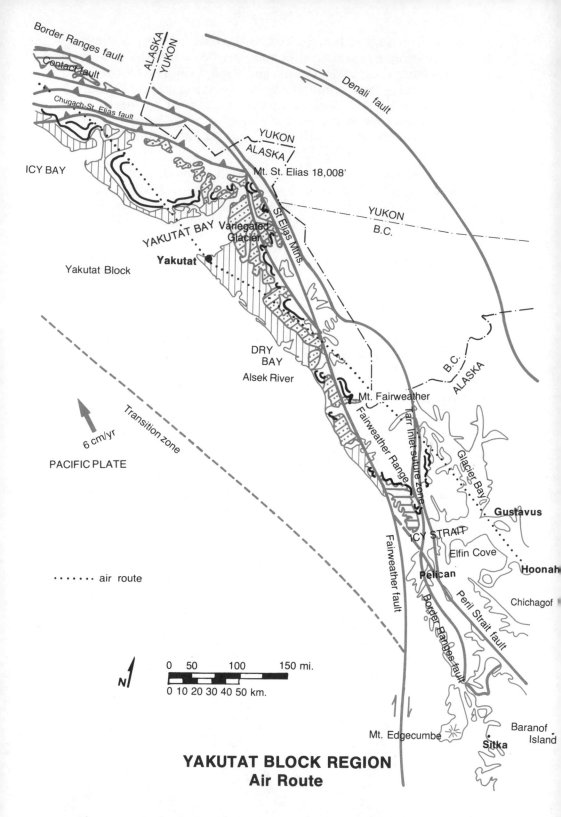

Border Ranges fault

Contact Fault

Chugach-St. Elias fault

ALASKA / YUKON

YUKON / ALASKA

ICY BAY

Mt. St. Elias 18,008'

YUKON / B.C.

Denali fault

St. Elias Mtns.

YAKUTAT BAY Variegated Glacier

Yakutat

Yakutat Block

DRY BAY

Alsek River

Mt. Fairweather

B.C. / ALASKA

Fairweather Range

Tarr Inlet suture zone

Glacier Bay

Gustavus

Transition zone

6 cm/yr

PACIFIC PLATE

ICY STRAIT

Elfin Cove

Hoonah

Fairweather fault

Pelican

Chichagof

· · · · · · · air route

Peril Strait fault

Border Ranges fault

0 50   100   150 mi.

0 10 20 30 40 50 km.

N

Mt. Edgecumbe

Baranof Island

Sitka

YAKUTAT BLOCK REGION
Air Route

Hubbard Glacier overrunning Osier Island enroute to blocking Russell Fjord northeast of Yakutat, view is to the north. April 1986.

—Rod Flynn photo

Yakutat's Surging Glaciers

By June of 1986 Hubbard Glacier, a giant river of ice at the head of Yakutat Bay, had surged hundreds of feet in just a few weeks and dammed Russell fjord. The huge lake formed south of the glacier was rising 4 inches daily. On June 18, 1986 the lake level was 6 feet higher than sea level on the north side of this galloping glacier. Continued blockage of the fjord caused the lake to rise to 83 feet nearing its 120 foot overflow level. Overflow would cause runoff from the newly formed 30 mile long, two mile wide deep lake to wash into the Situk River at the south end of Russell fjord sweeping away steelhead and salmon along with much debris. There is evidence that suggests the Situk Channel was flooded by Hubbard glacier surges between 200 and 500 years ago.

October 8, 1986 the Hubbard Glacier ruptured, dumping 3,500,000 cubic feet of water per second into Disenchantment Bay near Yakutat. This tremendous volume of turbulent water lifted deep water shrimp off the bottom of the fjord and threw them onshore. This may be the largest discharge in North America since late Pleistocene time, when glacial Lake Missoula emptied into the Columbia River.

Variegated Glacier is a small southern tributary of Hubbard Glacier, at the head of Yakutat Bay and opposite Gilbert Point in Russel fjord. It is 15 miles long, 1000 feet thick and caused a lot of excitement in 1983 when it surged. Alaska has 200 to 300 known surging glaciers that can advance up to 200 feet per day. Variegated Glacier surges about every 19 to 20 years.

A normal "slow moving" glacier makes its way down its valley by a combination of sliding at its base, and flow within the ice. Glaciers can thereby transfer ice from the upper glacier where snow accumulates to the lower glacier where it melts away. When a glacier completes this transfer evenly and ice removed downhill equals the ice accumulated in the upper glacier, it is said to be in "mass balance."

Glacial plumbing seems to be the key to surging. Surging glaciers are unable to move ice accumulated at their source areas to their lower ends on a regular basis. Every 20 years or so, exceptionally high runoff water from rain or spring thaw eases the glacier's movement. Then the weight of the excess ice accumulated in the upper part of the glacier forces rapid movement.

Surges are in some ways analogous to a good dose of Drano to a clogged sink. The surge seems to regulate the plumbing. Streams at the foot of surging glaciers can instantly burst into raging torrents, surprising unwary glacier enthusiasts. Why the surges occur about every two decades in Variegated Glacier is unknown. Surging glaciers are a problem in inhabited areas where they can create unstable ice dams across streams.

Malaspina Glacier, billed by airline pilot commentators as the glacier that is bigger than the state of Rhode Island (isn't everything in Alaska?), is the largest piedmont or "foot-of-the-mountain" glacier

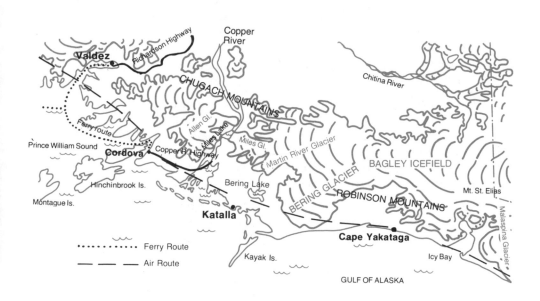

ICY BAY—VALDEZ
Ferry and Air Routes

Folded moraines on Malaspina Glacier west of Yakutat. –U.S. Geological Survey photo by D. J. Miller

in North America. It covers an area of about 850 square miles between Yakutat and Icy Bay and is supplied with ice by more than 25 tributary glaciers, most of which are in the Yukon. Studies measure ice thicknesses of between 1,130 and 2,050 feet. Looking out of the airplane window, you will see the strange series of deformed stripes along the perimeter of the glacier. These are moraines that have been contorted and twisted by iceflow within the glacier.

Like Glacier Bay, Icy Bay has come out from the cold only recently. When George Vancouver sailed by in 1794, Guyot Glacier filled Icy Bay. A small indentation at the bay's mouth had formed by 1837 when Sir Edward Belcher sailed along the coast, but it wasn't until 1904 that Guyot Glacier began its retreat to open up the bay. It is now about 25 miles long, which means a very large volume of ice has melted in the last 80 years. Since the last century, melting glacial ice has raised sea level by more than a foot.

Recent Glacial Activity along Alaska's South Coast

Glacial marine sediment younger than 10,000 years old is mostly fine-grained silt which has settled out of suspension. The greatest thickness is more than 975 feet just south of the Copper River. Other thick piles of glacial sediments are more than 750 feet thick offshore from Icy Bay, more than 450 feet thick in the Bering Trough south of the Bering Glacier, and more then 675 feet thick in the Hinchinbrook sea valley near the entrance to Prince William Sound. Glacial end moraines exist at the mouths of Lituya Bay, Icy Bay, and Yakutat Bay, at Fairweather Glacier, and possibly south of Bering Glacier.

During Pleistocene time, glacial ice may have reached beyond the edge of the continental shelf as a vast sheet of floating ice. Large volumes of sediment now found in deep-sea deposits as far as 900 miles south of the Alaskan coast were carried by floating icebergs. This evidence suggests that floating glaciers reached far south many times in the past, from at least 5 million years ago to the end of the last Ice Age about 10,000 years ago.

During the last 10,000 years, tongues of ice reached offshore in the Gulf of Alaska. Hubbard Glacier near Yakutat had its maximum Little Ice Age advance about 1000 years ago. Glaciers adjacent to Lituya Bay had their maximum Little Ice Age advance during the last 600 years. Glacially scoured troughs on the continental shelf offshore from La Perouse, Fairweather, and Grand Plateau Glaciers provide evidence for possible advances of these glaciers onto the shelf during the past 10,000 years. East of Yakutat, the Alsek River glaciers may have moved onto the continental shelf as recently as 4500-5000 years ago. Farther east the Glacier Bay–Cross Sound ice sheet left a submarine moraine in Cross Sound Sea Valley.

Tertiary marine and non-marine rocks of the Yakataga formation are glaciated by Guyot and Tyndall glaciers along the Gulf of Alaska.
—U.S. Geological Survey photo by D.J. Miller

Copper River Delta Region

As the plane heads west to Cordova on the edge of Prince William Sound, a vast expanse of tidal flats appears to the north. This is the complex delta system of the Copper River, the only river other than the Alsek near Yakutat that flows through the south central Alaskan coastal mountains. This old river flowed to the coast before the Chugach Mountains were slowly raised across its path. Tectonic forces, tidal action, and glaciation all combine to control construction of this delta. The Copper River's flow is greatest between April and September, peaking in May and June. From fall to early spring a high pressure area over the interior land mass funnels air seaward through the Copper River Canyon, causing very strong winds that airplane pilots have clocked at 96 miles per hour. The winds carry enough silt to build dunes on the upper delta.

During major Pleistocene glaciation, ice dammed the Copper River in the Chugach Mountains. Drainage then shifted to threshold areas across the Alaska range into the Yukon River and through the Talkeetna Mountains into Cook Inlet. The present drainage probably resumed flowing through the pre-Ice Age channel about 9000 years ago, after melting ice allowed drainage of Glacial Lake Atna in the Copper River Basin.

The inner continental shelf adjacent to the Copper River delta rose around the end of Pleistocene time, but has been slowly subsiding at a rate of about 0.1 to 0.5 inches annually for the last 1,700 years. Subsidence has been interrupted by occasional uplifts, which enable forests to grow and uplift took place during the 1964 earthquake, which also caused slumping, faulting, and sand volcanoes. The front of the Copper River delta slumped, spreading a great deal of sediment into deeper water. The same earthquake generated great sea waves that caused widespread erosion along the coast, and deposition in deeper waters.

No moose lived in the prime habitat of the Copper River delta until 1949. Coastal glaciers kept them from migrating from the unglaciated interior of Alaska. Between 1949 and 1953, 14 moose calves and one yearling were transplanted from the Kenai, Susitna, and the upper Copper River areas to the Copper River delta. Those moose would have eventually penetrated the Chugach Mountains to the Copper River delta on their own, but it might have taken them much longer.

The Copper River and Northwestern Railway

Between 1885 and 1888 Miles Glacier advanced to within 360 feet of the Million Dollar Bridge, built as part of the Copper River and Northwestern Railway line that was the crucial link between the seaport at Cordova and the copper mines at McCarthy. Miles Lake formed after the retreat of the glacier between 1884 and 1898. In the summer of 1910 Childs Glacier temporarily advanced to within 1425 feet of the bridge. Had that happened in winter when undercutting of the toe does not slow it, the glacier would have destroyed the bridge

A. The $1,500,000 Railway Bridge which was menaced by the advance of Childs Glacier in 1910. —U.S. Geological Survey photo

B. Part of front of Childs Glacier in Copper River showing the comparison of ice cliff, which varies from 250 to 300 feet in height, with the Capitol, in Washington, which is 287 feet high.
–U.S. Geological Survey photo

102

The proximity of glaciers to the river causes unpredictable violent floods when the glaciers suddenly release stored water. On February 8, 1909 an ice marginal lake in Miles Glacier discharged, raising river level at the Million Dollar Bridge from 18 to 36 feet in six days. After having survived flooding and glacial advance the bridge was destroyed in the 1964 earthquake.

Katalla

The first discovery of oil in Alaska was by placer miners who found seeps on the eastern Copper River delta at the former town of Katalla. The first oil claims were staked in 1896 and Alaska's first producing well was drilled in 1902 by a British firm that hit an oil producing formation 260 feet below the surface. Total production in the Katalla field amounted to 154,000 barrels by 1937, when a fire destroyed part of the refinery. The oil-bearing rocks at Katalla are within the belt of Tertiary marine sedimentary rocks lying along the Gulf of Alaska Coast between Katalla and Yakutat. Beds representing each of the epochs from Eocene to Pliocene occur in a belt 300 miles long and as much as 40 miles wide. It was the only oil field in Alaska until the Swanson River field was discovered in 1957 in Cook Inlet.

Cordova

On July 12, 1984 an earthquake registering 6.2 on the Richter Scale cracked open an ice dam east of Cordova, flooding thousands of acres on the Copper River delta, washing away trumpeter swan nests, and threatening salmon spawning grounds. The earthquake rattled Anchorage buildings and knocked food off the grocery shelves in Valdez. Two days later a U.S. Forest Service Ranger spotted a geyser of water escaping from the surface below the ice dam in Bering Glacier below Berg Lake, 60 miles east of Cordova. The level of the lake dropped 210 feet stopping several days later when the level was down to about 60 feet. The Gandil River, bank full with glacial lake runoff, flooded and covered the country around Bering Lake. The flood also covered territory under consideration as a transportation corridor from Cordova to coalfields near the Bering River.

From Cordova it is possible to travel across Prince William Sound by ferry to the town of Valdez, deep water seaport and terminus of the Trans Alaska Pipeline. Kayakers should watch out for supertankers which set up a pretty good sized wake.

Map Symbols for Eastern Alaska, the Yukon, and Northern British Columbia

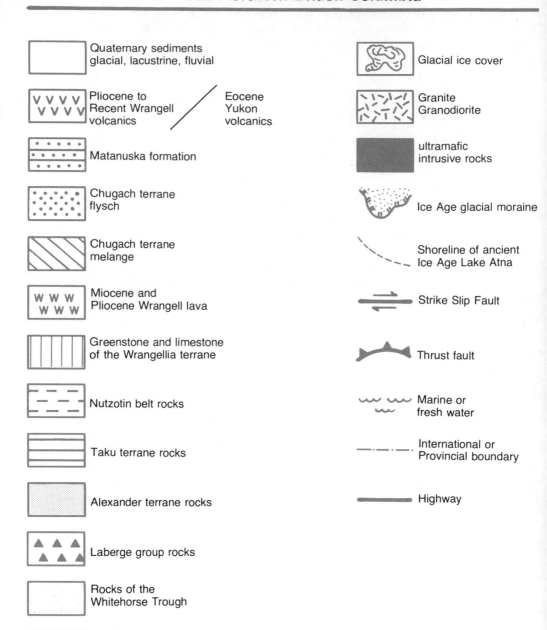

Quaternary sediments glacial, lacustrine, fluvial

Pliocene to Recent Wrangell volcanics

Eocene Yukon volcanics

Matanuska formation

Chugach terrane flysch

Chugach terrane melange

Miocene and Pliocene Wrangell lava

Greenstone and limestone of the Wrangellia terrane

Nutzotin belt rocks

Taku terrane rocks

Alexander terrane rocks

Laberge group rocks

Rocks of the Whitehorse Trough

Yukon-Tanana terrane rocks

Glacial ice cover

Granite Granodiorite

ultramafic intrusive rocks

Ice Age glacial moraine

Shoreline of ancient Ice Age Lake Atna

Strike Slip Fault

Thrust fault

Marine or fresh water

International or Provincial boundary

Highway

II
EASTERN ALASKA AND
ADJACENT YUKON

Haines—Haines Junction—
The Haines Highway
(see map accompanying Juneau—Haines)
158 mi./253 km.

The Chilkat Peninsula forms an important geological boundary in northern Lynn Canal. It is composed of Mesozoic greentstones, volcanic sandstones, mudstone, chert, and limestone that closely resemble rocks of the Gravina belt. These Gravina belt-type rocks lie on top of much older lower to middle Paleozoic carbonates of the Alexander terrane. Southwest of Haines, Alexander terrane rocks crop out in the the Chilkoot Mountains and farther southwest into Glacier Park.

Chatham Strait Fault

Along the Chilkat River Valley, Mesozoic Gravina-like rocks and Alexander terrane rocks are separated by the Chatham Strait fault, which can be traced from Berner's Bay north of Juneau, along Lynn Canal, and northwestward through the Chilkat River Valley. Poorly preserved fossil leaves exist in conglomerates and slates of Paleocene age in the Chilkat Valley. The fossils are along the Chatham Strait fault near the confluence of the Takhin and Chilkat rivers and were smeared out of shape by shearing movement along the fault sometime between about 40 million and 10,000 years ago.

Geologic features younger than 10,000 years have not been offset along the Chatham Strait fault. Studies made by marine geologists during 14 ship crossings of the fault in Lynn Canal, showed no sign of offset in the fjord-filling sediments deposited since the glacier melted back at the end of the last Ice Age. However in November 1987 an earthquake registering 5.3, epicentered near Haines, rattled southeast Alaska. The northern Chatham Strait fault is still active.

Ultramafic Rocks in Southeast Alaska

Cretaceous ultramafic rocks, intrusives composed entirely of dark minerals, are conspicuous near Battery Point south of Port Chilkoot, and near Klukwan northwest of Haines up the Chilkat River. They form a long belt that can be traced as far south as Duke Island at the southern end of the southeast Alaska panhandle. Some of these intrusions are zoned with an inner core of dunite, an uncommon rock composed almost entirely of green olivine. Outer shells of peridotite, pyroxenite, hornblendite, and gabbro surround the dunite cores. Such zoned complexes are rare; the only other place they are known to abound is in the Ural Mountains of Russia. Geologists believe these intrusions may form as dark iron and magnesium-rich minerals stuck to the walls of magma chambers that once fed overlying andesite volcanoes.

Bald Eagles Take Advantage of Local Geology

Near the delta of the Tsirku and Chilkat Rivers by the village of Klukwan, 750 feet of river sand and gravels fill this deep, glacially scoured valley. They were dumped there since the end of the last Ice Age. These sediments form a thick aquifer, or water-bearing layer, that discharges groundwater into the Chilkat river at a temperature of 40 degrees F., and prevents ice from forming in several sections of the river. Late runs of coho (silver) and chum (dog) salmon beginning in October, supply the river with a valuable winter food source. As many as 3,500 bald eagles congregate to take advantage of this concentrated protein along a three to five mile stretch of the Chilkat River until January when they finally head south for the winter.

The union of these two rivers created a natural settling spot for magnetite grains in the Klukwan fan. Their source is ultramafic rock that forms the steep canyon slopes of Iron Mountain. This 5,745 foot peak northeast of Klukwan, contains iron-bearing outcrops in an area about one mile wide and three miles long. Tributary streams carry the grains downstream to the Klukwan fan, which contains an estimated 800 million tons of iron ore that averages 10 percent magnetite.

The Haines Highway turns west out of Klukwan, to follow the Klehini River valley. The Chatham Strait fault, actually a segment of the Denali fault, trends northwestward from Klukwan along the Kelsall River Valley.

Porcupine Gold

The Old Porcupine Mine and ghost town, about 35 miles up the road from Haines, lie across the Klehini River at its confluence with Porcupine Creek. Gold was discovered in 1898 in stream placer deposits about two miles above the mouth of the steam. The deposits were mined until 1906 when a flood on Porcupine Creek destroyed the flumes built to separate the heavy gold from the lighter sands and gravels. Small placer operations are still active. Between 1898 and 1955, 60,000 troy ounces of gold were extracted from the Porcupine Creek placers. Recently gold was discovered near Mount Henry Clay at the headwaters of Glacier Creek, southwest of Porcupine.

Copper from the Rainy Hollow District

The Rainy Hollow mining district lies at the head of the Klehini River Valley just north of the British Columbia boundary. Loose chunks of copper ore were first found at Rainy Hollow by miners enroute to the Klondike along the Dalton Trail. The copper mineralization is on Mineral Mountain and Copper Butte. Small veins reaching from a diorite pluton are mineralized along the contact of Paleozoic marbles with mudstone, quartzite, gneiss, and schist. The ore minerals are bornite, chalcocite, chalcopyrite, dark brown to black sphalerite, and galena.

Many claims were staked in 1898 and later years. The Maid of Erin Mine is at the 3,600 foot level on the southwestern slope of Mineral Mountain west of the hairpin turn at mile 52 on the Haines Highway. Between 1911 and 1922 this mine produced 77,658 pounds of copper, 5,849 ounces of silver, and 6 ounces of gold from 157 tons of sorted ore.

The 300-mile Dalton Trail connected Haines with the gold fields in the Yukon near Dawson City. This trail was longer but less grueling than the Chilkoot Trail out of Skagway. Traces of the original trail and several of Jack Dalton's caches or "roadhouses" remain along the Klehini River.

Chilkat Pass

About 60 miles northwest of Haines, the Haines Highway crosses Chilkat Pass at an elevation of 3,493 feet. Chilkat Pass is an important geographical divide that separates the Alsek ranges and St. Elias Mountain block to the south and west from the Coast Mountains to the north and east. It was also an important point along an age-old trade route that was controlled by Tlingit Indians of the Chilkat region. Later, as gold seekers swarmed into the region, Jack Dalton laid out a 300 mile toll road that connected Haines and the Yukon along the Chilkat Pass route.

North of the pass, Kelsall Lake lies on the east side of the highway northeast of Mule Creek and the Glacier Camp highway maintenance station. The Chilkat segment of the Denali fault lies between the Haines Highway and Kelsall Lake.

Rise of the Saint Elias Mountains

The Haines Highway passes through the Tatshenshini River Valley for about 14 miles. It contains a deep fill of glacial sediments. The valley separates the Alsek ranges from the Boundary ranges of the Coast Mountains on the east. This river, as well as others that flow parallel to the Haines Highway, has entrenched its course in a northwest-trending plateau called the Duke Depression. This feature extends along the southwestern side of the Kluane Range.

The youngest and highest mountains of Canada began to rise after late Miocene time, about 12 million years ago. Before this, the Wrangell Lavas, erupted in late Miocene time, flowed over a surface of low relief. Interbedded with these lavas are sediments and coal deposits that formed in basins at low elevations. After the Yakutat block jammed against the southern coast of Alaska, the Saint Elias Mountains began to rise. The forces that raised the mountains fractured the Wrangell Lavas, and fragments were lifted high above the original flow surface. Now isolated outcrops of the Wrangell Lavas can be found on some of the highest ridges and peaks in the Saint Elias Mountains.

The Denali fault, which forms the northern boundary of the Saint Elias Mountain block, shuffled the rocks further. Muddy sandstones deposited during Jurassic and Cretaceous time from the source north of the Denali fault between Klukshu and Kluane Lake have been offset at least 180 miles southeast of matching rocks south of the Denali fault in the Nutzotin Mountains of Alaska. Given time, the Denali fault will move Mt. Logan northwestward from the Yukon into Alaska, perhaps forcing future Canadian park rangers to wear Alaska uniforms.

At the border between British Columbia and the Yukon, eskers west of the road record streams that flowed in or beneath glaciers. They form parallel clusters of gravel ridges 50 to 100 feet high and as much as two miles long.

Dezadeash and Kathleen Lakes

Between about 1.8 and 5 million years ago, a single large stream probably carved the valley now occupied by the Takhini and Dezadeash rivers. This drainage once carried water southwest into Alsek River, possibly through the valley now occupied by Dezadeash Lake. That lake now drains to the north into the Dezadeash River, which flows through a narrow valley near Mt. Bratnober, then turns west to enter the Alsek River. The floor of Dezadeash Lake is now 3,000 feet higher than the Takhini River Valley and is cut off from it by the Denali fault, which runs between them along the Shakwak Trench. The Dezadeash Lake Valley and southwest side of the fault have risen 3,000 feet relative to the northeast, Takhini River side of the fault. Drainage was disrupted with the continued rise of the Saint Elias Mountains during late Pliocene time.

During the Pleistocene Ice Ages, Lowell Glacier flowed from the St. Elias Range into the Shakwak Valley through the present site of Kathleen Lake. The glacially carved basin was later dammed by sand and gravel deposited at the mouth of Victoria Creek to form the Kathleen Lakes.

Peridotite, containing massive serpentine with fibrous asbestos, occurs at an elevation of 2,750 feet at the mouth of the Quill River, 2½ miles north and west of the Kathleen River. The peridotite formed originally deep in the earth's mantle beneath the ocean crust. The asbestos mineral is chrysotile and fibers up to 2 inches in length have been found. Asbestos is an excellent fireproofing material and was widely used in building construction during the middle of this century. Recent findings that the tiny fibers cause lung cancer has lead to the removal of asbestos from many buildings.

Just south of Haines Junction, the highway crosses the Dezadeash River, part of the Alsek River system that flows into the Pacific Ocean near Yakutat. The Alsek and Copper rivers are the only two that completely transected the coastal mountains to enter the Gulf of Alaska.

Kluane National Park

Kluane National Park, in the southwest corner of the Yukon, contains the highest mountain in Canada, 19,850 feet Mt. Logan, and the largest non-polar icefield in the world. In 1980 Kluane and neighboring Wrangell St. Elias National Park in Alaska were jointly commemorated as a United Nations World Heritage Site. Together they are recognized and protected under the UNESCO World Heritage Convention as an outstanding wilderness area of global significance.

More than half of Kluane is covered by mile-deep ice fields, which are nourished by frequent snowstorms originating to the south in the Gulf of Alaska. Canada's largest and longest valley glaciers, the Donjek, Kaskawulsh, and Lowell, flow out in all directions, to the lower and outer edges of the park. Cirques, moraines, glacially gouged valleys, hanging valleys, and rock glaciers punctuate the park's glaciated landscape.

The southern Tuchone Indians were the original inhabitants of this region. Europeans came later in search of gold and in 1903 over 1,000 miners rushed to the Kluane gold fields. Ruins of some of the gold workings can be seen in Sheep and Bullion creeks.

In the northeast corner of Kluane Park, near the Alaska Highway, sediments of Williscroft Creek contained a 6,000 year-old fossil *Bison (Superbison) crassicornis*. At an earlier time, the southern Yukon provided good grazing for these extinct grass eaters. At Kluane Park visitor center in Haines Junction, a seismograph records earthquake activity in the region.

Aerial view of Ogive Glacier near Skagway. The glacier is named for the curved bands of debris-laden ice that are convex downslope.
—U.S. Geological Survey photo by A. Post

110

Skagway—Whitehorse—
Klondike Highway 2
(see map accompanying Juneau—Haines)

Fossils in the Coast Range

From the southern Yukon south for approximately 1800 miles,
through southeastern Alaska to Vancouver, British Columbia, the
coastal mountains are made up of four parallel rock zones, mapped
according to metamorphic or igneous type. This region is known as the
Coast Range Metamorphic Plutonic complex. Cretaceous fossils occur
in the westernmost zone and Permian and older fossils have been
found in the easternmost zone. The two central zones contain generally metamorphosed rocks that were not known to contain fossils until
1984.

About 30 miles southeast of Skagway near the south end of Taku
Arm in Tagish Lake new mapping by U.S. Geological Survey geologists led to a discovery of lenses and beds of relatively unmetamor-
phosed limestones that contain certain fossil ammonites. These relatives of today's deep-sea dwelling, chambered nautilus became extinct
at the end of Cretaceous time. The fossils near Skagway are from
middle Triassic time, 200 to 220 million years ago. They are important
because their host rocks are similar to rocks in Juneau that contain
very few recognizable fossils. The ammonite-bearing rocks strongly
suggest that rocks in the Juneau area, once considered part of the
unique "Tracy Arm terrane", are more likely a metamorphosed
portion of the Alexander terrane.

Ultramafic Rocks Near Skagway

About 4 miles southeast of Skagway, along the northeastern edge
of the Juneau Icefield, is a newly discovered occurrence of periodotite.
This massive body of rock composed of olivine, clinopyroxene, and
magnetite, is about 18 miles long and 5 miles wide. It intrudes gneiss,
gabbro, and amphibolite. The peridotite is not related to other southeast Alaskan periodotites distributed between Klukwan near Haines

south to Duke Island near the the southern tip of Prince of Wales Island. Such ultramafic rocks have source areas in the earth's upper mantle. These iron- and magnesium-rich rocks may mark the sites of ancient volcanic island arcs and help unravel the geologic history of the region.

Skagway and the Chilkoot Trail

Granitic rocks crop out abundantly in and around Skagway. Here Cretaceous diorite and Cretaceous to early Tertiary granodiorite intruded older Mesozoic rocks. The older rocks were raised during emplacement of the intrusions; later erosion, which included the bulldozing action of Ice Age glaciers, removed them leaving the granitic rocks exposed in mountain peaks and roadcuts.

Two miles north of Skagway is the junction with the road to Dyea. Outcrops of diorite along the roadway and along the nearby mountainsides show bedrock that is grayish white, massive, jointed, and polished by ice. Some boulders of diorite show a very weathered envelope or are disintegrating to form sand as their iron minerals oxidize to rust leaving piles of feldspar behind. In many places the bedrock is capped by windblown glacial silt perhaps carried by the famous north wind from which Skagway derives its name. By 1898 the discoveries of Klondike gold inspired nearly 2000 prospectors to struggle over an old Chilkoot Indian trading route to reach the goldfields near Dawson in the northwestern Yukon. From the tidewater boomtown of Dyea near Skagway, goldseekers climbed 3,739 feet in 16.6 miles to cross the coastal mountains northward into the Yukon River drainage basin.

On the Alaska side, this well-travelled trail follows the narrow Taiya River valley past Irene glacier. The valley is fringed by many 6,000 foot peaks of Cretaceous to early Tertiary granodiorite and culminates in the infamous scree slope known as "The Scales". Here treacherous piles of loose granodiorite talus provide the only footing over the Chilkoot Pass into Canada.

The upper end of the Chilkoot Trail and the summit at Chilkoot Pass in 1898. Men, sleds, animals, and supplies are bound for the Klondike gold fields.

—U.S. Geological Survey

photo by A.H. Brooks

Men easing sleds down the east side of Chilkoot Pass into British Columbia from Alaska, April 1898. —U.S. Geological Survey photo by A.H. Brooks

The landscape changes remarkably north of the pass. The Canadian side of the trail features a rocky plateau of wide open country, newly emerged from its Pleistocene ice cover. It is made of ice-rounded mountain peaks flanked by bare rocky slopes with sparse vegetation. Many small icefields drain into deep, glacially-carved lake basins. The coastal mountains act as a rain barrier, confining the 200 inches or so of annual snowfall to the summit areas. These are treacherous places when the northwind blows.

White Pass

Although relatively short, the Chilkoot Trail was hard on pack animals. Affluent prospectors opted to buy horses, goats, dogs, mules, oxen or even camels and took the White Pass route over the Coast Mountains into the Yukon. Its summit of 3,290 feet make it 439 feet lower than the Chilkoot Pass. The trail to White Pass is more gently graded than the Chilkoot but it does have a steep approach to the summit.

In 1900 the White Pass and Yukon Railroad was completed, enabling anyone capable of buying a train ticket to make the White Pass route. Between Skagway and the summit are numerous freshly blasted highway exposures of Cretaceous diorite. Zones of green chlorite can be seen in road cuts near the pass. At the summit, outcrops of pale Tertiary granite reveal a younger intrusion into the older diorite.

Between Skagway and White Pass, Klondike Highway 2, completed in 1978, is very steep, and challenges the brakes of heavily laden ore trucks now transporting lead and zinc ore from the Cyprus Anvil mine in Faro, Yukon, to the tidewater port at Skagway. The highway caused the closure of the White Pass and Yukon Railroad, which made its last run between Skagway and Whitehorse in 1982. In 1990 a short stretch of the old run was reopened for summer use between Skagway and Log Cabin north of White Pass.

113

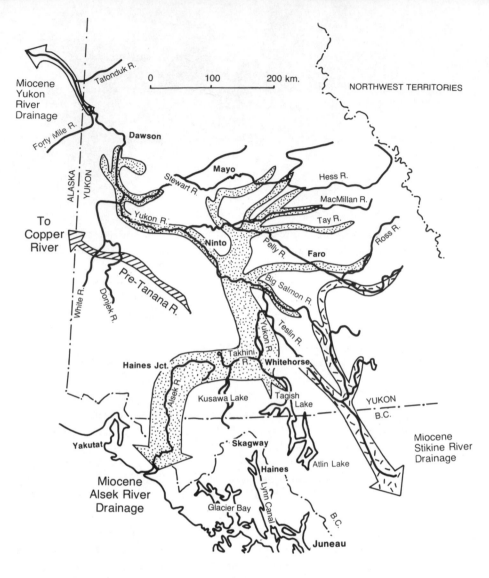

Miocene drainage patterns of the Yukon River.

The New Drainage Basin
of the Yukon River

Once across White Pass, Klondike Highway 2 enters the westerly draining Yukon River watershed. Between about 22 and 5 million years ago, in Miocene time, the southern and central Yukon Territory drained southward into the Gulf of Alaska, through the ancestral Alsek and Stikine rivers.

Waters now flowing into the Yukon River did not have the barrier of the St. Elias and Coast Mountains during early and middle Miocene time. By middle Miocene time the Yakutat block, which probably began as a slice of the southeast Alaska continental margin, moved northwestward along the Queen Charlotte-Fairweather transform fault system and began colliding directly with the southern Alaska coast. The St. Elias and Coast Mountains then rose unevenly above interior parts of the Yukon. This new tilt to the landscape drastically changed the pattern of rivers flowing across it.

During Pleistocene time, drainage patterns were rearranged further by ice. The headwaters of the Yukon River system were near Dawson about 2 million years ago. As glaciers and ice caps melted beginning 12,000 to 10,000 years ago, the St. Elias Mountains were slower to thaw than regions farther north in the interior. The entrenched tributaries of the Alsek and Stikine rivers were unable to carve new channels southward through ice-choked mountain canyons. One by one they were captured by the quickly expanding Yukon River drainage, and forced to abandon their 180-mile route to the Gulf of Alaska and flow along the much less direct route 1500 miles to the northwest, across interior Alaska into the Bering Sea.

As the lower part of the Stikine River near Wrangell became ice-free, it was able to reclaim some of its tributaries from the Yukon River. What had been the headwaters of the ancient Alsek River during Miocene time became the source of the Yukon River at the end of Pleistocene time. This was a great boon to Klondike goldseekers who, having survived the perils of the Chilkoot Trail or White Pass, were able to float their supplies down Lake Bennett, cross at Carcross to Tagish and Marsh Lakes, and descend the Yukon River through Whitehorse to Dawson and the goldfields.

This may be a temporary situation. The Yukon River is constantly in danger of losing its newly acquired tributaries to more energetic streams that flow directly into the Gulf of Alaska. Perhaps a million years from now, Yukon River canoeing routes will more closely resemble those of Miocene time, and far distant Alsek River kayakers will be able to put their boats in at Lake Bennett.

The Carcross Region

Between Log Cabin and Tutshi Lake, Klondike Highway 2 crosses the boundary between the intrusive granitic rocks of the Coast range and the older Triassic country rocks they intruded. This region of rocks is called the Whitehorse Trough. Rocks along the shore of the Tutshi Lake are upper Triassic andesites and basalts that are part of the Stuhini Group, a group of volcanic and sedimentary rocks that formed during Mezosoic time as a chain of volcanic islands.

Gold and Silver Mining in the Windy Arm Area

Just north of the border between British Columbia and the Yukon is the old Venus Mine stamp mill. The highway follows the eastern flank of Montana Mountain, which has a noticeably reddish appearance. Reddish or altered rocks guided prospectors to zones of mineralization. Gold and silver were removed from the Venus Mine from an adit on the eastern slope of Montana Mountain above the Windy Arm of Tagish Lake and transported to the stamp mill by an aerial tramway. The Arctic Caribou (or Big Thing) Mine, also worked for gold and silver, is on Sugar Loaf Hill on the north side of the Montana Mountain. The Windy Arm area deposits have been worked off and on since 1905.

Montana Mountain contains a diverse group of rocks. As Klondike Highway 2 rounds Windy Arm from south to north it crosses first the rocks of the lower Jurassic Laberge group. They include muddy sandstones, arkose, conglomerate, graphitic phyllite, quartzite, and greenstone. Near where the highway crosses the border between British Columbia and the Yukon are the much younger Eocene Sloko rhyolites and trachytes. Gold and silver from the Venus Mine came from veins in the Sloko volcanic rocks.

The Atlin Terrane

A few miles farther north, the rocks again change, and road cuts expose the much older Carboniferous and Permian Anvil range group—andesites, basalts, cherts, and tuff. They are in the northern edge of the Atlin terrane, a region of upper Paleozoic deep-water sediments and dark volcanic rocks faulted against Mesozoic rocks of the Whitehorse Trough. Atlin terrane rocks were intensely folded during their uplift in late Jurassic time.

As the highway rounds the bend near Bove Island to turn northwest toward Carcross, the rocks change abruptly to the Cretaceous to Tertiary granodiorite found on the upper portions of the Chilkoot Trail. Bove Island and Lime Mountain, across Tagish Lake to the east, are Carboniferous to Permian Horsefeed limestone, also part of the late Paleozoic Atlin terrane.

Carcross

Bouldery silt that looks like glacial till exposed in roadcuts near Carcross reminds roadside geologists that glacial ice occupied this region during Pleistocene time. The highway passes northwestward toward Carcross by Nares Lake and Lake Bennett. During Miocene time, this area was headwaters for the Alsek River, which now begins near Dezadeash and empties into the Gulf of Alaska near Yakutat.

Carcross stands on an old caribou migration route that passed through the narrows between Lakes Bennett and Tagish. Lakeshore visitors in Carcross are commonly blasted by fine sand carried in the winds blowing off Lake Bennett. This sand was originally dumped into a glacial lake that occupied the entire area. North of town, in the direction of Whitehorse, is an accumulation area for windblown glacial sands. The presence of such a huge sand pile provides evidence for the strength and persistence of these local winds.

Between Carcross and the intersection of Klondike Highway 2 with the Alcan Highway, outcrops are few and far between. A fault just north of Carcross separates the upper Paleozoic carbonate and volcanic rocks of the Atlin terrane from the Laberge group rocks of the Whitehorse Trough. Although poorly exposed here, these Jurassic sediments are prominent in roadcuts along the shores of Tutshi Lake to the south.

Miles Canyon

About 35 miles south of Whitehorse is Miles Canyon and the Yukon River, an impressive sight that is well-recorded on postcards and travel books throughout the greater Whitehorse region. The Pliocene to Pleistocene Miles Canyon basalts contracted while cooling to form polygonal-sided columns; they were later covered by Pleistocene glacial and river deposits. The Yukon River created a beautiful geologic cross section as it carved this 3,000 feet long by 90 feet wide channel. The river drops 16 feet through the gorge and once provided white water thrills to boat travelers enroute to Whitehorse.

Schwatka Lake

Whitehorse Rapids, which once separated the gold miners from the not so lucky and gave the Yukon's capital city its name, are now at rest beneath Schwatka Lake. This dammed section of the Yukon River generates hydro power for Whitehorse and vicinity. The lake is named for Lieutenant Frederick Schwatka who made the first rough survey of the Yukon River under orders from the U.S. Government. The Canadian authorities were not notified of the expedition and were understandably perturbed when Schwatka, ignorant of their names, renamed most of the rivers, lakes, and prominent hills he encountered. Schwatka later attempted to climb Mount St. Elias, but his 250 pound frame prevented success.

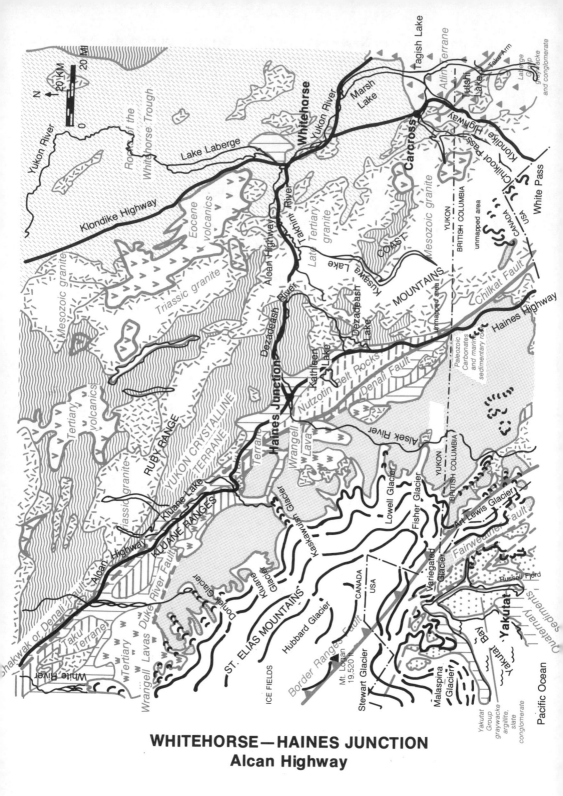

WHITEHORSE—HAINES JUNCTION
Alcan Highway

THE ALCAN

Whitehorse—Haines Junction
97 mi./155 km.

Whitehorse

Miners and travelers taking the river trip from Lake Bennett through Whitehorse Rapids in the late 1800s used Whitehorse as a resting point. It became the upstream launching point for steamers headed down the Yukon River to Dawson City; they returned to the "United States" by continuing downriver to St. Michael on the Bering Sea coast. Whitehorse became the terminus for the White Pass and Yukon Railroad in 1900. Forty-two years later the Alaska Highway, built during World War II, linked Whitehorse by road with the rest of Canada.

In Riverdale, a small Whitehorse suburb on the east side of the Yukon River, the Plio-Pleistocene Miles Canyon lavas outcrop beside the fish ladder just below the dam. The basalts are riddled with vesicles, holes made by gas bubbles escaping while the rock was still molten. The vesicles later filled with mineral deposits carried by ground water circulating through the basalt. Vesicle fillings include radiating tufts of aragonite crystals and white grape-like clusters of calcite. Also present are dog-tooth calcite crystals and brownish-yellow olivine.

A Recently Rearranged Yukon River

Overlying the basalt are deposits of silt and boulder clay; the remnants of the pre-glacial Yukon River. The Cordilleran ice sheet moving from the east and the St. Elias ice sheet, moving from the southwest, overlapped in the vicinity of Whitehorse during glacial advances of Pleistocene time. Glacial ice dammed the Yukon and other rivers, created temporary. lakes, and caused a major rearrangement of the drainage patterns. In the Whitehorse river bluffs on

119

either side of the Yukon River, there are glacial lake deposits of finely layered silt. Until 1.8 million years ago, the upper Yukon River took a shortcut to the sea by following a 189 mile route through the Takhini-Dezadeash valley, out to the Alsek River, and through the coastal mountains into the Gulf of Alaska near Yakutat. The Alsek River kept pace with the rising St. Elias Mountains during Pliocene time, cutting its channel through the rapidly rising coastal barrier. As the St. Elias ice sheet formed in the mountains during Pleistocene time, it filled the Alsek River valley and blocked river access to the sea. Ice remained in the mountains long after it had melted in the lowland interior near Whitehorse. The lower Yukon River pirated the flow from drainages that were trapped north of the St. Elias ice, and the rivers in the Whitehorse vicinity were forced to flow into the Yukon River drainage. This lengthened their route to the sea by about 1300 miles.

The Whitehorse Copper Belt

The broad valley of the Yukon River is about 4 miles wide at the Whitehorse and is bordered on the east by a long ridge of upper Triassic limestone known as Grey Mountain. This limestone is part of the Lewes River group, a sequence of Mesozoic volcanic and sedimentary rocks that underlie the Whitehorse trough. Farther west, Cretaceous to Tertiary granodiorites intruded older Lewes River group rocks. The molten granitic rocks reacted with the limestone to form skarn minerals and deposit copper minerals.

Copper mineralization in the Whitehorse area occurs along an 18½ mile zone between Porter Creek at the north end and to Cowley Lakes to the southeast. Many different types of ore minerals have been found, including chalcopyrite, bornite, and native copper. Other metallic minerals include pyrite, magnetite, specular hematite, molybdenite, pyrrhotite, arsenopyrite, stibnite, galena, and sphalerite. Non-metallic minerals such as epidote, garnet, diopside, plagioclase, actinolite-tremolite, and calcite can be found along the skarn zone. Copper-bearing outcrops were first noted in 1897 by gold-seeking Klondike miners traveling north to Dawson and the gold fields there. Ore was later shipped south via the White Pass and Yukon Railway. Rocks of the Lewes River group crop out in roadcuts between Whitehorse and Fish Lake Road, which provides access into the copper region.

Geology Along the Takhini River

About 10 miles north of Whitehorse, the Alaska Highway turns west, leaving the moderate Yukon River valley and entering the

Takhini River valley. Mesozoic granodiorites of the Boundary ranges of the Coast Mountains form the peaks to the south. On the north side of the highway are the late Tertiary quartz monzonites of the Miner's Range. The Alaska Highway follows the old Whitehorse-Kluane wagon road built in 1904. The Dawson Trail followed the west side of the Miners Range.

The Takhini River and Kusawa Lake once drained into the Alsek river and emptied into the Gulf of Alaska near Dry Bay. At present they are Yukon River tributaries and flow an extra 1200 miles through interior Alaska to enter salt water along the Bering Sea coast.

More than 10,000 years ago, an Ice Age glacial lake collected the layered silts that now crop out along the highway cut and stream channels. The Alcan Highway follows the old lake bottom through this area. The much younger White River volcanic ash, was erupted far to the southwest less than 2,000 years ago, appears above the silts in noticeable white layers.

About 40 miles east of Haines Junction the highway passes through Champagne, an old resting point along the Dalton Trail. A retreated glacier melted back toward its source area and left the prominent ridge made of intermixed boulder, gravel and sand-sized materials. Modern winds have created dunes on top of the moraine by mantling it with layers of fine sand and silts.

The Shakwak Valley and Denali Fault

About 8 miles from Haines Junction, Kluane Park and the St. Elias Mountains come into view. The highway enters the Shakwak Valley and passes through it from the next 150 miles to the White River. The bedrock changes from Mesozoic granodiorite of the coast ranges to Jurassic schist, gneiss, and amphibolite of the Kluane formation.

This unusual valley lies along the Shakwak segment of the Denali fault. It is oriented along a northwesterly trend from the westernmost point of Kusawa Lake to and beyond the White River. The Denali fault separates the geologically younger St. Elias Mountains to the south from the older rocks of the Yukon Plateau on its north side.

There is good evidence that the Shakwak Valley formed by vertical fault movements rather than the horizontal style of movement along the Denali fault. The very straight southwest side of the valley is 1300 to 1500 feet above the northeast valley wall. The walls of the Shakwak Valley are fault scarps that bound a block that dropped during Pliocene time. The fault on the southwestern side of the valley continued moving after the northeastern side had stopped, enabling modern streams to cut about 300 feet more deeply into the northeastern valley wall. The Shakwak Valley was next mantled with Pleistocene and recent glacial and volcanic sediments.

Movement along this segment of the Denali fault may have started as early as Cretaceous time. Rocks of the Wrangellia terrane in the Kluane ranges were strongly deformed during Cretaceous time, forming tight folds and metamorphic banding. The metamorphic bands were later kinked by compressional forces. Studies of fractured cobbles in Tertiary conglomerates show that the stresses occurred between late Oligocene and early Miocene time, about 15 to 30 million years ago, and were probably related to faulting.

Haines Junction—Tok
302 mi./483 km.

Kluane Mountains

Between Haines Junction and the White River, on the southside of the Alcan Highway, the Kluane Range marks the northern boundary of the St. Elias Mountains. The Kluane Range is composed mainly of the eastern Skolai terrane that was intruded by Jurassic and Cretaceous granodiorite magma. Skolai terrane volcanic rocks are upper Paleozoic, mainly Permian, andesites and basalts with some sedimentary rocks. These rocks are also found in the eastern Alaska Range and the Wrangell Mountains, where they closely resemble rocks of the Taku terrane in southeastern Alaska. The volcanic rocks outcrop in the central Alaska Range and in the Kluane Range and appear to be the remnants of a volcanic island chain formed directly on oceanic crust during late Paleozoic time.

Ruby Range

Bordering the north side of the Alcan Highway from Aishihik River to the Kluane River, the Ruby Range is underlain by Triassic granodiorite and Jurassic-Cretaceous biotite schist, gneiss, and amphibolite. Some of the Ruby Range streams have been worked for placer gold. In both the Ruby and Kluane ranges, trees cease to grow at elevations above 4,000 feet.

Seven Miles west of Haines Junction, the 15 mile trail to Sugden Creek begins on the south side of the highway just west of the bridge over Bear Creek. About eight miles down the trail, outcrops of peridotite lie in an intrusion along the mountain slope on the west side of the Dezadeash River. Here mineral collectors can find crystals of green olivine measuring 3 to 4 inches in length as well as light green tabular crystals of diopside. Magnetite particles within the olivine give a blackish cast to its green color. At the end of this trail the Sugden Creek placer deposits have been worked for gold and platinum. Peridotite commonly contains platinum, so it is no surprise to find such placers in this stream.

Near Kloo Lake and Jarvis Creek, a series of morainal ridges straddling the highway were left by ice sheets that moved northeast during Pleistocene time.

Boutillier District

About 35 miles northwest of Haines Junction, northbound travelers get their first glimpse of Kluane Lake at Boutillier Summit, which at 3,280 feet is the highest point on the highway between Whitehorse and Fairbanks. Fossil plant pollen has been located from Ice Age sediments in Silver Creek at the south end of Kluane Lake. It is between 30,000 and 65,000 years old. The pollen helps geologists to recreate ancient plant communities and figure out what Ice Age mammals, such as woolly mammoths, ate. The Boutillier fossil pollen shows that plants like those that now grow at higher elevations in the Boutillier District, once grew near Kluane Lake. They included sedges, grasses, sage, willow, and a few trees. This pollen information tells geologists that the Shakwak Valley was ice free between 30,000 and 65,000 years ago and that the climate was drier and colder than it now is.

Kluane Lake

Kluane Lake is 40 miles long and between 2 to 6 miles wide, the largest lake in the Yukon and the deepest in the Shakwak Valley. It lies at an elevation of 2,575 feet and drains through the Kluane, Donjek, White, and Yukon rivers into the Bering Sea.

Near the mouth of Silver Creek, the discovery of placer gold led to the construction of a settlement known as Kluane or Silver City. It later became the center of the Kluane Mining District. Gold was first discovered 15 miles northeast of Kluane Lake on Fourth of July Creek in the Ruby Range. A gold rush followed after Dawson Charlie of Carcross staked his claim on July 4, 1903. Gold has been mined in the Kluane district off and on ever since this discovery. Most of the gold was found in placers south of the present highway. They were in Sheep, Bullion, Burwash, and Arch creeks and the Koidern River. Very few new deposits have been found since 1905.

Slim River

Fierce winds funnel fine-grained silt, derived from the Kaskawulsh Glacier, down the Slim River Valley where it is carried into Kluane Lake. A large river delta at the south end of the lake has grown from the constant deposition of river-carried silt into the lake, creating raised mudflats and altering shorelines.

Roadcuts along this stretch of the Alcan Highway are made of Upper Triassic Nikolai greenstone, an altered basalt that contains copper. It is probably the source of copper in the McCarthy District of Alaska and in the White River District of the Yukon. The copper fills gas cavities and fractures in the basalt and was probably emplaced during metamorphism at low temperature, possibly beneath the seafloor.

Along the shoreline of Kluane Lake and in the bed of Congdon Creek are pebbles of red chert, called jasper. They are especially numerous at the Goose Bay campsite, and probably originate in the Permian Skolai terrane rocks.

Duke River

On the south side of the highway is a wide gap in the Kluane ranges, occupied by the valleys of the Duke River and Burwash Creek. Through the gap is flat-capped Amphitheater Mountain, conspicuous in the foreground of the Donjek Range. The flat cap is Tertiary Wrangell basalt that overlies sedimentary strata. Coal seams and fossil leaves are found in light brown and grey shale along the slopes below the lava cap. These rocks and fossils provide important time markers that help to date the rise of the St. Elias Range and arrival of the Yakutat block.

Donjek River Valley

Northwest of Burwash landing the highway follows the Kluane River. North of the highway, morainal ridges of sand and gravel lie between the Kluane and Donjek rivers. The Donjek River, marks the western boundary of the Donjek ranges, and drains from the Donjek and Steele glaciers. Steele Glacier is known locally as the Galloping Glacier because it surged 1600 feet in one month in 1966-67.

The Donjek River carries large volumes of gravel, sand, and silt through constantly shifting channels that wander across the valley floor. Geodes or thunder eggs filled with chalcedony and quartz can be found in the river gravels. In Pleistocene time the valley was filled with glacial ice.

White River Region— Copper and Volcanic Ash

On the east side of the White River a short, single lane road south of the highway, leads to the Canalask Mine overlooking Miners Ridge and the Nutzotin Mountains. Ore minerals of copper, lead, zinc, and nickel occur in altered volcanic rocks as small lenses of massive ore.

The Canyon City copper deposit is accessible from the Canalask Road. Large slabs of native copper have been found here. One weighed 2,590 pounds. It is now in the MacBride Museum in Whitehorse. Copper slabs are thought to have weathered out of fractures in the Triassic Nikolai basalt, where copper also fills gas bubbles and small fractures.

Local Indians made cooking utensils and weapons from nuggets they dug out of the rock with caribou horns. It was also valuable in trade for fish oil and salmon with the Tlingit people south of the coastal mountains.

Cliffs of Wrangell andesite and glacial outwash rise above gravel benches along the White River opposite Travor Creek, Alaska near the Yukon border. —U.S. Geological Survey photo by S.R. Capps

White volcanic ash forms a widespread layer a few inches to two feet thick throughout the southern Yukon. It commonly mantles lower mountain slopes, but also occurs as a thin layer beneath topsoil, and as a thin band in roadcuts and cut banks of streams and rivers. The ash is made of pumice that looks like white sand; some fragments of pumice as much as 4 inches long have been found.

A volcanic explosion that happened less than 2000 years ago near the Natazhat Glacier, about 30 -miles southwest of the White River Bridge, was probably the source of the ash. Dunes of ash several hundred feet high have been found there. During the eruption, winds carried the ash into the Yukon and spread it across two lobe-shaped areas: one to the north along the Yukon–Alaska border into the Ogilvie Mountains, the other to the east into the MacKenzie Mountains. The thickness of ash deposits generally decreases away from the source area. The fairly even distribution of the White River Ash suggests it may have fallen steadily for several days over a wide area.

Nutzotin Mountains

At the bridge over Sanpete Creek, the Alaska Highway leaves the Shakwak Valley, which continues northwesterly into the Nutzotin Mountains and into Alaska. The highway rejoins the Yukon Plateau, which occupies a wide area extending as far north as Dawson City.

About seven miles from the Yukon–Alaska border, a broad area north of the highway is free of glacial deposits. The Pleistocene icesheets that originated in the Icefield ranges south of the Kluane range moved northward from Shakwak Valley and reached their northern limit in this area.

The highway now follows the flank of the Nutzotin Mountains to the southwest. The eastern end of the Alaska Range forms two principal mountain masses: the Mentasta and Nutzotin Mountains, both composed entirely of rocks of the Nutzotin belt. These Jurassic and lower Cretaceous marine sedimentary and volcanic rocks extend from the eastern Alaska Range to an eastern boundary against the Denali fault in the Yukon. Isolated outcrops of Nutzotin belt rocks known as the Dezadeash group appear intermittently between Dezadeash Lake and the border between the Yukon and Alaska. Near Dezadeash Lake, there are extensive exposures of slate, graywacke, argillite, quartzite, chert, impure limestone, grit, conglomerate, tuffaceous sandstone, and bedded volcanic tuff. Dezadeash rocks also occur in small patches to the northwest in the Kluane Lake area within the Kluane ranges. Some geologists believe these similarities provided evidence to link the Nutzotin belt rocks with those of the Gravina belt in southeastern Alaska.

In the eastern Alaska Range, two distinct groups of rocks make up the Nutzotin belt. The Nutzotin Mountains sequence, the oldest is a variety of shallow and deep water deposits. Thick sections consist of gray mudstone, siltstone, and muddy sandstone. Fossils are rare. A few late Jurassic clams have been found between Chisana and the Yukon border. The upper part of the Nutzotin belt is the Chisana formation, a thick sequence of andesitic rocks that erupted both beneath the sea and on land. They are exposed near Nabesna and throughout the Chisana area.

The main source of Nutzotin sequence sediments was the Yukon terrane, as indicated by the pebbles in conglomerate layers. A rapidly sinking basin accumulated the Nutzotin belt rocks in middle Jurassic through late Jurassic time, about 175 to 140 million years ago. Sometime between about 150 and 100 million years ago, rocks in the Wrangell Mountains were folded, uplifted, and deeply eroded. Volcanic eruptions that produced the andesite of the Chisana formation were probably caused by subduction in the Chugach trench before the Wrangellia terrane docked.

Yukon-Tanana Terrane

The region north of the Denali fault contains metamorphic rocks that include phyllite, quartz-mica schist, and marble. Late Mesozoic and early Tertiary granitic intrusions are locally abundant. Low temperature greenschist metamorphism is common in Yukon–Tanana terrane rocks in south-central Alaska and the Yukon. In Alaska, the rocks may grade to the north into higher temperature metamorphic rocks known as Birch Creek schist. Late Mesozoic rocks of the Nutzotin belt do not appear to have been deposited upon the rocks of the Yukon–Tanana terrane, even though pebbles in the Nutzotin belt came from this terrane. In the eastern Alaska Range and in the Kluane Range, the Denali fault separates the two rock packages.

Denali Fault

From the International border to Northway Junction, the Alaska Highway follows the Denali fault; its trace through this stretch lies south of the highway. This part of the fault appears as a discontinuous, north-facing bluff in young sedimentary deposits. Glacial moraines from the last Ice Age have been horizontally offset as much as 753 feet. The Denali fault sweeps a broad arc from the Queen Charlotte fault system in southeastern Alaska, across the Canadian border, back into Alaska through the Alaska range, and westward to Bristol Bay on the Bering Sea coast.

The Totschunda fault trends northwest from the Yukon border for 120 miles to its junction with the Denali fault near Mentasta Pass. For most of its length this fault can be traced by conspicuous topographic features such as low scarps, ponds, fissures, offset drainages in bedrock, and offset glacial deposits. Displacement of glacial features along the Denali and Totschunda faults yields estimates of fault movement averaging one quarter of an inch per year during the last 30,000 years. Displacement probably happened before the White River ash erupted about 1600 years ago because soils containing the ash are not offset.

Tanana River

Along this stretch of highway the Chisana and Nabesna rivers flow north, and join to form the Tanana River near Tetlin Junction and the turnoff for the Taylor Highway. The Chisana River, Chisana Glacier, and the Nabesna River show offset of 2.2 miles across the Denali fault. With displacement rates of one half to one inch per year, movement on these fault segments could have begun as recently as 200,000 to 400,000 years ago.

Before Pleistocene time, the Pliocene Tanana River Valley was probably occupied by a single stream that flowed northwest into the Pacific by way of the Copper River. During the Pleistocene time, the White River, deflected by an ice dam near the Yukon border, cut through the north wall of the Tanana Valley. That forced the White River into its present course into the Yukon River north of the Dawson range. The Tanana River now follows the northeastern edge of the Alaska range, and enters the Yukon River northwest of Fairbanks.

Tok—Delta Junction—
Fairbanks
64 mi./103 km.
94 mi./151 km.

The Alcan Highway follows the Tanana River all of the way from Tok to Fairbanks. Between Tok and Delta Junction, the road goes through hills of Precambrian schist intruded by Cretaceous granite. West of Delta Junction, the north side of the road follows the same hills of schist, part of the Yukon-Tanana terrane. South of the road, sand and gravel in an outwash plain reach to the Alaska range on the southern skyline. On a clear day, visible mountains from east to west are Mounts Hays (13,832 feet), Hess (11,940 feet) and Deborah (12,339 feet).

Yukon-Tanana Terrane

Bedrock in this area is metamorphic rock of the Yukon-Tanana terrane, which contains granitic intrusions. Once known collectively as the Birch Creek schist, the metamorphic rocks are mainly quartz-mica schist, micaceous quartzite, and graphitic schist. This is probably a composite terrane, a collection of pieces of crust that were metamorphosed together. Their individual histories will be very difficult to sort out.

The Yukon-Tanana terrane may contain the oldest rocks in Alaska. Uranium-lead isotope data from zircons eroded from older rocks indicate that the metamorphic rocks are recrystallized sedimentary rocks which were deposited about 2.2 billion years ago. The sediments were probably deposited at what was then the edge of the continent, some hundreds of miles to the south and east, then faulted into their present location.

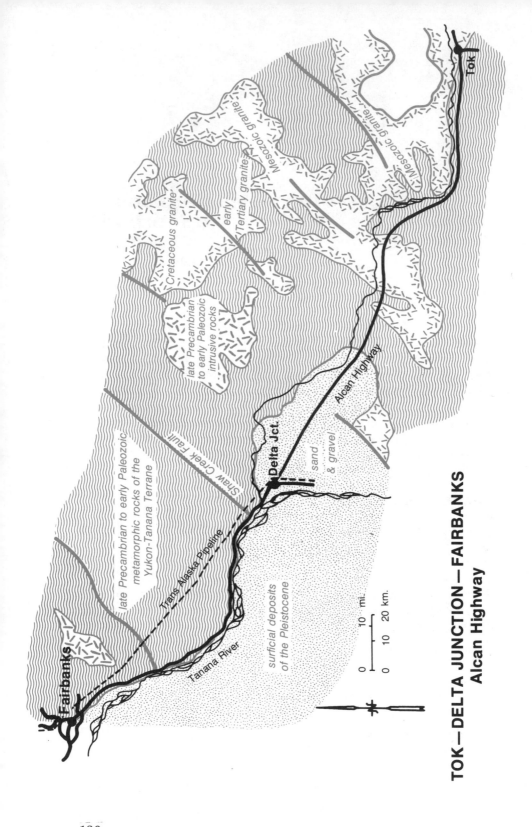

TOK—DELTA JUNCTION—FAIRBANKS
Alcan Highway

Fairbanks

Tanana River

Trans Alaska Pipeline

late Precambrian to early Paleozoic metamorphic rocks of the Yukon-Tanana Terrane

Shaw Creek Fault

late Precambrian to early Paleozoic intrusive rocks

Cretaceous granite

early Tertiary granite

Mesozoic granite

Delta Jct.

sand & gravel

surficial deposits of the Pleistocene

Alcan Highway

Tok

0   10 mi.

0   10   20 km.

Zircons are fairly common as tiny crystals in granitic igneous rocks. The mineral typically contains a measurable quantity of uranium and so is useful for age dates based on the rate of radioactive decay. Other physical properties of the mineral make it especially useful. Zircon is very hard, so zircon crystals, weathered out of granitic rocks, removed by erosion, and then redeposited as sediment, generally survive in fairly good shape. From them it is possible to know the age of the original rocks from which the metamorphic rocks were born. Another physical property of zircon that makes the procedure reliable is its high melting point. Once a mineral is brought near its melting point, its atoms are scrambled, isotopes mixed, and the radioactive clock reset. It is fortunate that zircon has a high melting point because the rocks of the Yukon-Tanana terrane may have been subjected to half a dozen episodes of metamorphism during their long history.

The first two-thirds of the way from Tok to Delta Junction, large masses of Cretaceous granite intruded the metamorphic rocks. Cathedral Bluffs, 25 miles west of Tok and across the river from the highway, and Tower Bluffs farther west on the north side of the river, are made of granite.

The first few miles of highway west of Delta Junction are on an outwash plain of sand and gravel with large, rounded cobbles. The Tanana and its tributaries carry a heavy load of sediment. Their braided character is typical of streams with a steep gradient and a heavy load of gravel. Farther west, the river is caught between glacial deposits on the south and hills of metamorphic rocks on the north.

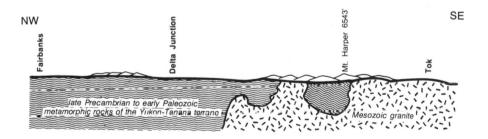

Section through the Yukon-Tanana terrane between Fairbanks and Tok. The granite is much younger.

131

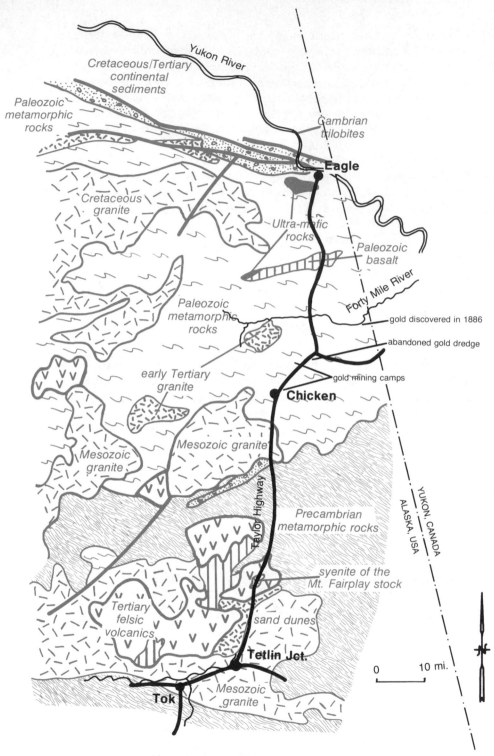

Yukon River

Cretaceous/Tertiary
continental
sediments

Paleozoic
metamorphic
rocks

Cambrian
trilobites

Eagle

Cretaceous
granite

Ultra-mafic
rocks

Paleozoic
basalt

Forty Mile River

Paleozoic
metamorphic
rocks

gold discovered in 1886

abandoned gold dredge

early Tertiary
granite

gold mining camps

Chicken

Mesozoic granite

Mesozoic
granite

Taylor Highway

Precambrian
metamorphic rocks

syenite of the
Mt. Fairplay stock

Tertiary
felsic
volcanics

sand dunes

YUKON, CANADA
ALASKA, USA

Tetlin Jct.

0   10 mi.

Tok

Mesozoic
granite

N

TETLIN JUNCTION—EAGLE
Taylor Highway

Sand dunes are a common feature in post Ice Age Alaska. These dunes lie beside the road five miles north of Tetlin Junction. —DO

Tetlin Junction—Eagle
137 mi./219 km.

The Yukon-Tanana Upland

The Taylor Highway winds north through rolling hills of the Yukon Tanana upland, a complex of metamorphic rocks that are some of the oldest in Alaska, metamorphosed at least 600 million years ago. The older rocks were intruded by granite, and covered with lava and sediments. Rocks exposed along the way include a broad spectrum of igneous and metamorphic rocks and continental sedimentary rocks.

Sand Dunes

For about seven miles north of Tetlin Junction, the road goes through a field of sand dunes. The sand is mostly dark gray and medium-grained. It blew off the floodplains of the Tanana River and other rivers. In summer, the wind still blows sand and silt off the river bars and barren glacial till.

133

Intrusive and Volcanic Rocks

For about four miles south of Mt. Fairplay, the road goes through granite and syenite of the Mt. Fairplay stock, a small body of coarse-grained igneous rock. Syenite is a widespread, but uncommon igneous rock like granite but dominated by orthoclase and lacking igneous rock quartz. Recent highway construction left good specimens along the road. The syenite is dark gray, coarse-grained, and contains large crystals of orthoclase feldspar.

The light greenish and reddish rocks exposed along Logging Cabin Creek north of Mt. Fairplay are weathered volcanic debris erupted during early Tertiary time, around fifty million years ago. Rhyolite lavas are typically charged with steam, and erupt explosively. The volcanoes that produced these lavas eroded long ago, but may once have resembeld the modern ones in Katmai National Park.

Between Taylor Creek and Chicken, are good exposures of the deeply weathered Taylor Mountain batholith. In the fashion of granite, the rock of the batholith has crumbled into loose debris. Age dates place the age of the granite in Jurassic time, about 150 million years ago. The igneous rock varies in composition, primarily through variations in the proportions of quartz and feldspars. Some geologists find these distinctions interesting, but most would lump all the rock types together under the general heading of granite.

Gold

Between Chicken and Eagle, the road again crosses metamorphic rocks of the Yukon-Tanana terrane. These rocks formed through metamorphism of shale, siltstone, sandstone, and limestone into schist and gneiss, quartzite and marble. Later, small mineralized quartz veins were precipitated near the contacts of the metamorphic rocks with granite intrusions; the quartz veins were the source of gold in the placers of the Fortymile mining district in the area of Chicken and Eagle. Gold has been mined in this region since its discovery on the Fortymile River in 1886 and soon thereafter at Chicken, Jack Wade, and near Eagle. The rush was a prelude to the slightly later, and larger, gold rushes to the Klondike, Nome, and Fairbanks.

There is still gold in these hills, but nearly all of the valuable placers have been staked and worked at least once. Every summer brings more prospecting and mining. A modern operation with bulldozer, front-end loader, and sluice box has been operating next to the road in recent years along Wade Creek. In 1963 a 25 ounce nugget was found. Floating suction dredges operate in the summer on the Fortymile River.

134

In this modern gold placer mine at Jack Wade Creek, a front end loader puts gravel into a sluice box where gold collects in the riffles. —DO

Tertiary conglomerate, sandstone, and siltstone underlie the heavily wooded area south of Eagle. These continental sediments contain coal and plant fossils. They were deposited about fifty million years ago in a basin forming on the interior side of the westward-moving Yukon-Tanana terrane.

About four to seven miles south of Eagle, the roadcuts expose a slice, of ultramafic igneous rock. Such rocks normally exist in the earth's mantle. These rocks are typically dark green when fresh, and weather to a reddish brown. They lack quartz and other silica-rich minerals, and are dominated by dense, iron- and magnesium-rich minerals like pyroxene, olivine, and serpentine. The latter comes from the chemical reaction of water with the olivine. Specimens of these minerals and asbestos, a fibrous variety of serpentine, can be found along American Creek.

The town of Eagle stands precariously perched on the eroding banks of the Yukon River. It had its beginnings in 1874 when a trading post was established there. The town boomed for a while after the 1886 discovery of gold on the Fortymile River and until the turn of the century it was the civil, military, and communications center for the Interior. The population dwindled after miners and prospectors were lured away by gold finds in Nome and Fairbanks.

Map Symbols for South Central Alaska

 Glacial Ice

 Quaternary sediments
glacial, greenstone, fluvial

 Prince William terrane

 flysch of the
Chugach terrane

 melange of the
Chugach terrane

 Paleozoic & Mesozoic
sedimentary rocks

 Paleozoic & Mesozoic
sedimentary rocks

 schist & gneiss

 Lavas

 Pillow Basalts

 Granite

 Diorite

 Ultramafics

 Glacial Moraine

 drumlins

 thrust fault

 Strike Slip Fault

 marine or
fresh water

 Highway

III
SOUTH-CENTRAL ALASKA
ANCHORAGE AND THE
KENAI PENINSULA

THE GREATER ANCHORAGE AREA

Anchorage is not far removed from the ice ages. At least five times in the last two to three million years, great glaciers have moved down into Cook Inlet and the Anchorage bowl. The first three times, ice filled the inlet and there was 2000 feet of ice on the Anchorage area. The latest couple of glacial episodes saw great glaciers advancing into the Anchorage area, but not completely filling Cook Inlet.

Bootlegger Cove Formation

During the last Ice Age, the site of Anchorage was at the intersection of glaciers moving down Turnagain and Knik Arms of Cook Inlet. Glaciers emptied into Cook Inlet and then, scattered glacial outwash and till across the lowlands as they melted back. Fine-grained sediments settling out of the milky water at the glaciers' edge accumulated to form the notorious Bootlegger Cove formation that underlies much of the Anchorage area. The formation is peppered with sand that melted out of dirty ice floating in the water. As the great weight of the glacial ice melted and ran back into the sea, the land rose.

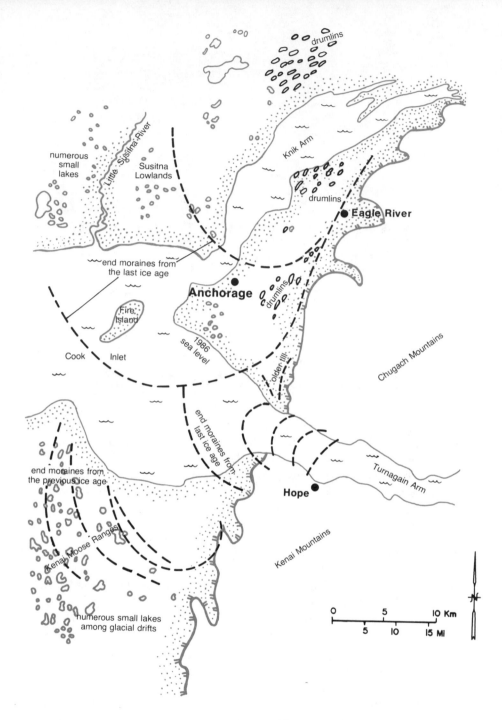

numerous
small
lakes

drumlins

Susitna
Lowlands

Little Susitna River

Knik Arm

drumlins

● Eagle River

end moraines from
the last ice age

Anchorage ●

drumlins

Fire
Island

1986
sea level

older till

Chugach Mountains

Cook   Inlet

end moraines from
last ice age

Turnagain Arm

end moraines from
the previous ice age

Hope ●

Kenai Moose Ranges

Kenai Mountains

numerous small lakes
among glacial drifts

0   5   10 Km

5   10   15 Mi

N

Moraines record the history of glaciation in the Anchorage area.

The Turnagain Heights landslide in 1964 followed the earthquake.

—U.S. Geological Survey photo

The Bootlegger Cove formation is mostly silt with up to about 5 percent clay minerals. It is mainly very fine-grained quartz and feldspar, rock flour formed by glacial abrasion. The formation is notorious because it slid so disastrously during the 1964 earthquake. Most of the landslides moved where the Bootlegger Cove formation is exposed in steep slopes along ocean bluffs and stream banks.

Good Friday Earthquake

The Good Friday earthquake of 1964 had a magnitude of 8.4, which places it among the great earthquakes of modern times. Anchorage, the nearest large city, was hit very hard. Parts of town along the coast were especially hard hit where the Bootlegger Cove formation slid into the sea. In the Turnagain area, 75 homes were totally destroyed, some sliding as far as 2000 feet. The damage was greatest in and around Earthquake Park. Large cracks appeared in the earth behind the slides, as far back as Northern Lights Boulevard.

Government Hill, one of the oldest parts of town, was heavily damaged. The Government Hill Elementary School was flattened; fortunately school was not in session. The citizens of Anchorage later decided to turn the land into a park rather than rebuild the school.

139

The collapse of Fourth Avenue in downtown Anchorage due to an earthquake-triggered landslide. —U.S. Geological Survey photo, 1964

Downtown Anchorage was devastated. Portions of 3rd and 4th Avenues dropped as much as 10 feet as the ground gave way and slid toward the sea. After the earthquake, many tons of slump debris were removed from the Buttress Zone and replaced with rock in an effort to hold back the landslide. This zone of massive failure is the grassy slope in Quayana Park between downtown and the railroad station.

The Bootlegger Cove formation, the source of all the slides, consists largely of glacial silt deposited on the bottom of the shallow estuary that formed in Cook Inlet when the ice retreated. Glacial lakes formed from meltwater impounded behind large glaciers that flowed down Turnagain and Knik arms of Cook Inlet. Violent ground shaking during the earthquake caused the glacial sediments to collapse and slide into the inlet.

The lessons of 1964 already seem to have been forgotten. Tall buildings have gone up downtown along the edge of the slide zone. Elsewhere, development proceeds on and next to landslides that moved during the earthquake. After the quake, those who lost their homes were given building sites in another part of town at nominal cost, but the state failed to buy the landslide areas, and now they are being developed again. Some of the more slide-prone land commands a higher price because it lies on a slope, and so has a good view.

Glacial Erratics

Glacial erratics are common around town. These are large boulders that have been transported from far away and dropped by a glacier or an iceberg. They are called erratics because they differ from the bedrock beneath them. Sometimes a small rock protruding at the surface will grow as it is dug out until it presents itself as a house-sized monolith much too big to move without the help of dynamite.

Peat and Ash

Before settlement, Anchorage was even boggier than it is now. Wet soils and deposits of peat exist here and there around town. Normal construction procedure is to remove the peat and replace it with sand and gravel before building. The Sears Mall, at Northern Lights Boulevard and the Seward Highway, is an example of a site where this has been done. Anchorage is now running out of sand and gravel, so fill material is being hauled by train from Palmer.

Layers of volcanic ash are visible in the peat, recording the history of ash eruptions from local volcanoes. Within the past generation, Anchorage received dustings from Spurr and Augustine volcanoes.

Mt. Susitna

The rounded mountain in the distance west of Anchorage is Susitna, the 'sleeping lady.' It is made of quartz diorite, a pluton of intermediate composition intruded into the Peninsular terrane during Jurassic time. The mountain owes its rounded shape to glacial ice that flowed across the summit. Erratic boulders left from an early glaciation litter its summit. At that time, the ice in Cook Inlet must have been at least three thousand feet thick.

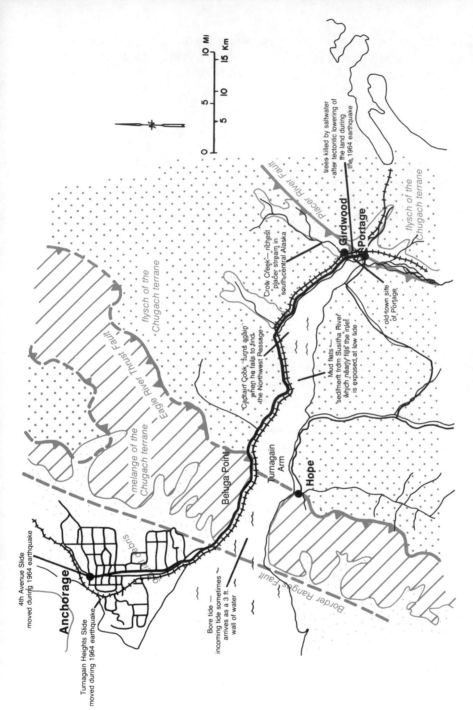

trees killed by saltwater after tectonic lowering of the land during the 1964 earthquake

Placer River Fault

flysch of the Chugach terrane

Girdwood

Portage

Crow Creek — richest placer stream in south-central Alaska

flysch of the Chugach terrane

old-town site of Portage

flysch of the Chugach terrane

Eagle River Thrust Fault

melange of the Chugach terrane

"Captain Cook 'turns again' when he fails to find the Northwest Passage"

Mud flats — sediment from Susitna River which nearly fills the inlet is exposed at low tide

Beluga Point

Turnagain Arm

Hope

Border Ranges Fault

4th Avenue Slide moved during 1964 earthquake

Anchorage

Turnagain Heights Slide moved during 1964 earthquake

debris

Bore tide — incoming tide sometimes arrives as a 3 ft. wall of water

10 Mi

15 Km

ANCHORAGE—PORTAGE

142

Anchorage—Portage
48 mi./77 km.

Beluga Point Archaeological Site

Beluga Point at the entrance to Turnagain Arm was the home of prehistoric hunters. For at least 8,000 years hunters came to this strategic location to hunt beluga whales and mountain sheep. Pieces of bone and stone blades have been recovered from the loess that covers bedrock at the site. Carbon-14 dates from different artifact-bearing layers in the loess indicate that occupation began about 9,000 years ago, and continued until 800 years ago.

Chugach Terrane

The Chugach terrane is well exposed in roadcuts along most of Turnagain Arm. Here we see the eroded remnants of rocks that were swept into a trench in late Cretaceous time—65 to 100 million years ago. The Chugach terrane contains a belt of melange and a belt of impure sandstone or graywacke. The two belts are separated along Turnagain arm by the Eagle River thrust fault, which crosses the road near Indian Creek and continues across Turnagain Arm and up Resurrection Creek.

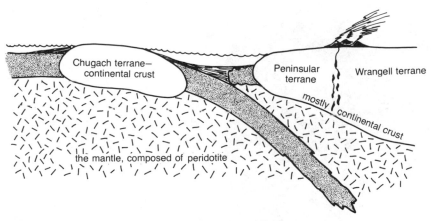

The arrival of the Chugach terrane about 65 million years ago along subduction zone in southern Alaska.

143

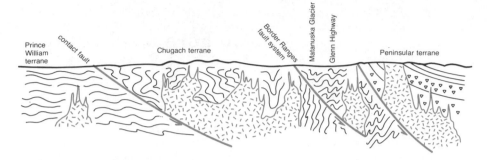

Geologic cross section through Chugach terrane. Faults are sutures that juxtapose dissimilar rock types of the different terranes. Granitic rocks intruded after suturing of terranes.

Melange is a French word meaning "the putting together of diverse things." In France, a salad with a little bit of everything thrown in would be a melange. Geologists borrowed this word to describe rocks like those exposed along the highway between Anchorage and Indian Creek. The McHugh Complex is a jumble of blocks and pieces of rock floating in a gray, fine-grained matrix. The blocks are of many different types of rock. Some are up to 25 miles long. The melange has been metamorphosed and sheared.

Flysch is a term that originated in Switzerland and refers to the collection of sediments shed by a mountain range, usually volcanic, as it is uplifted and eroded. Such sediments include mudstones, impure sandstones and conglomerates, all deposited in sea water. Between Indian Creek and Girdwood there is a fine collection of flysch rocks. The latest magnetic studies suggest that these rocks were

A melange is made of pieces of different types of rock that have been mixed together by movements of the earth's crust. —DO

144

originally deposited 25 degrees south of their present latitude, then drifted north on plates in the Pacific before being plastered onto the North American plate during the later part of the Cretaceous. They are referred to collectively as the Valdez group.

Turbidites can be seen in some of the outcrops between Indian Creek and Girdwood. Turbidites are deposited by turbidity currents, submarine slurries of sediment avalanching down the edge of a continental shelf into deep water. As a cloud of sediment comes to rest on the deep ocean floor, the larger particles come to rest first, creating graded bedding. By observation of the upward gradation from sand to shale in the bedding, it is possible to determine the original orientation of the sediments, even if the beds were later overturned.

The melange of the McHugh Complex and the flysch of the Valdez group are separated by the Eagle River thrust fault, which is well exposed along the highway immediately west of Indian Creek. The fault zone is several hundred yards wide. Look for breccia, angular fragments of chewed-up rock, and slickensides, a polished and grooved surface formed when two masses of rock move past each other under pressure.

A tidal bore entering Turnagain Arm, 1955.
—U.S. Geological Survey photo by D.R. Nichols

145

The Tidal Bore

Cook Inlet has one of the greatest tidal fluxes in the world, up to 28 feet. The tides in Turnagain Arm are impressive to watch and dangerous for boats. If you are lucky, you may be treated to the spectacle of a tidal bore, the first wave of an incoming tide. Tidal bores develop in shallow water where there is a large difference between low and high tide. A tidal bore is a wave analagous to a sonic boom. The wave travels fastest in deeper water, so the leading edge is curved, with the part in the middle of the inlet out in front.

Placer Gold

Crow Creek, north of Girdwood, has been the most productive placer gold stream in southcentral Alaska; over 40,000 ounces of gold since mining began in 1896. Every year recreational panners collect several tens of ounces. Much of the gold is coarse and nuggets are occasionally found.

These rails were bent by the force of the 1964 earthquake.
—U.S. Geological Survey photo

The old town of Portage was abandoned after the land was lowered by the 1964 earthquake and then flooded by the salt waters of Turnagain Arm. –DO

Good Friday Earthquake

During the Good Friday Earthquake, about 100,000 square miles of land was either raised or lowered, as a block of the earth's crust tilted. Portions of the Gulf of Alaska were raised as much as thirty feet; elsewhere the land was lowered as much as ten feet. After the earthquake the town of Portage had to be abandoned. The old buildings still stand opposite the Portage train stop, on the inlet side of the highway. The land sank enough there to allow salt water to flow in and kill the trees at the roots. The area now floods during high tide. Sediment deposited since the earthquake has begun to bury the buildings—you have to stoop to walk through the doorways. Across the highway, in the lower valley of the Twentymile River, tectonic lowering produced happier results. After the earthquake the land became marshy and full of ponds—good habitat for waterfowl.

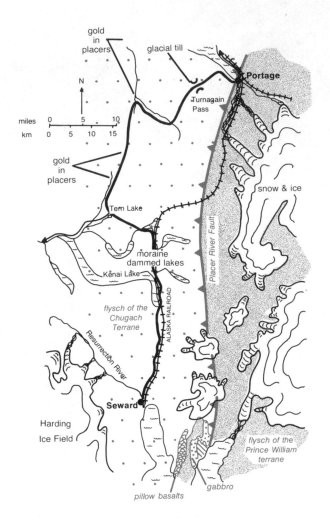

gold
in
placers

glacial till

Portage

N

Turnagain
Pass

miles 0 5 10

km 0 5 10 15

gold
in
placers

snow & ice

Tern Lake

Placer River Fault

moraine
dammed lakes

Kenai Lake

*flysch of the
Chugach
Terrane*

ALASKA RAILROAD

Resurrection River

Seward

Harding
Ice Field

*flysch of the
Prince William
terrane*

pillow basalts

gabbro

PORTAGE — SEWARD

The Harding Icefield on the Kenai Peninsula.
—National Park Service photo by M.W. Williams

Portage—Seward
81 mi./130 km.

Chugach Terrane

After leaving Turnagain Arm the highway climbs to Turnagain Pass and goes through the Kenai Mountains. The bedrock is metamorphosed muddy sandstones of the Chugach terrane. These rocks originated as rhythmically interbedded sandstones, mudstones and shales, deposited by turbidity currents and the slow settling of fine sediment in deep water. Later they were swept into a trench and metamorphosed. The total thickness of these deformed strata is unknown, but it must be at least a couple of miles.

Gold Mining

The road goes through and passes many small gold placers in part of the Hope mining district. Most of the gold comes from gold-bearing quartz veins that penetrate the metamorphic rocks of the Chugach terrane. Gilpatrick Ridge has many lode deposits. In lode deposits of this type, the microscopic gold is commonly not visible to the naked eye. An assay must be obtained to determine the gold content of the

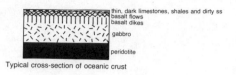

thin, dark limestones, shales and dirty ss
basalt flows
basalt dikes

gabbro

peridotite

Typical cross-section of oceanic crust

W

E

slice
of oceanic crust

Kenai Mountains

Harding Ice Field

Bear
Glacier

Resurrection Bay

pillow basalt

basalt dikes

gabbro

pods of
peridotite schist

Day Harbor

metamorphic rocks of
the Chugach terrane

metamorphic rocks of
the Prince William terrane

Cross section through Kenai Mountains and Prince William terrane. A slice of oceanic crust, an ophiolite section, was caught in the suture zone between the Chugach and Prince William terranes.

quartz veins. By the turn of the century, placer mining had begun on most of the streams that produced gold. Resurrection Creek was the most productive. There were small gold rushes at Sunrise and Hope. A couple of dozen small placer operations are now active on the Kenai Peninsula.

Turnagain Pass

Low mounds scattered on the flats south of the rest stop are glacial moraines. The topography is best seen in late afternoon or early morning when the light is low and shadows accentuate the relief. Some spectacular avalanche paths are on the east side of the road.

Harding Icefield

Exit Glacier got its name because it's about the easiest way to get off the Harding icefield without flying. A trail leads to the glacier, then parallels it up to the icefield. The icefield, 30 miles by 50 miles wide, is a remnant of the icecap that once completely covered the Kenai Mountains. Four glaciers, more than 15 miles long, still descend from the icefield.

The surface of the icefield is smooth and relatively flat and covers all but the tips of the higher peaks. A peak poking up through the ice is called a nunatak. The icefield is cold, wet, and subject to whiteout snowstorms on fifteen minutes notice, so be prepared to get weathered in if you go onto the surface. Bring skis.

150

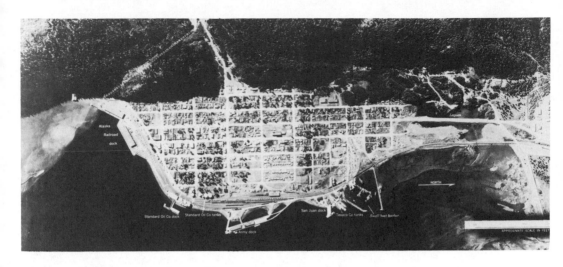

Seward before the 1964 earthquake. —U.S. Geological Survey photo

Seward one day after the earthquake of March 27, 1964. 30 seconds after the earthquake began, 4,000 feet of waterfront along Washington St. slid into Resurrection Bay. The waterfront industry was built on water-saturated alluvium similar to that underlying the town of Valdez. The earthquake caused the ground to lose it s shear strength and the docks, small boat harbor, and industries slipped away in mammoth underwater slides. Submarine slope adjacent to waterfront was 30 to 35 degrees. This damage was all prior to the arrival of the tsunamis! The tsunamis finished off what was left of the waterfront, destroying the Alaska Railroad docks and the State's southernmost railway terminus, an important sea port for Anchorage and the interior. Thin white sinuous line is probably the furthest inland reach of the tsunamis. —U.S. Geological Survey photo

Ophiolites

A partial ophiolite suite of rocks is exposed on the Resurrection Peninsula south of Seward, out of sight of the road. Ophiolite is the name given to a collection of rocks that normally makes up the ocean floor, a slice of the oceanic crust that has become part of a continent. This typically happens when plates crunch together, mixing the rocks up.

Rocks that compose an ophiolite suite are the same kinds you would encounter if you drilled a very deep hole through the floor of the ocean far from land. There the ocean floor is covered by a thin layer of very fine sediment, an ooze that accumulates very slowly as the tiny skeletons of planktonic plants and animals settle out of the water. Some plankton known as Radiolaria live in silica shells that slowly convert to chert after they are buried.

Beneath the layer of fine, siliceous sediments, the floor of the ocean is made of magesium- and iron-rich igneous rocks, basalts, that erupted from long cracks in the ocean floor. When basalt erupts underwater and cools, it contracts and develops a surface texture resembling a pile of rounded pillows. These basalts overlie their slower cooling, coarser-grained equivalent, gabbro. In some places, the feeder dikes of basalt are so abundant that they make up a "sheeted dike complex." Gabbro forms from the same magma as the basalt, but instead of being erupted on the sea floor, it is forced beneath the layers of basalt.

If you were to drill a very deep hole, you would eventually drill through the mafic oceanic crust and reach ultramafic rocks of the earth's mantle. Rock from the upper mantle, peridotite, is dark and very dense, and consists mostly of pyroxene and olivine. We sometimes see these rocks, along with chert and shale, on the continents, even though they originated in the oceanic crust. They are common as slivers caught between opposing blocks of crust. The olivine is easily altered to serpentine by the addition of water, and slides like grease along fault planes. At the surface, these rocks weather to a rusty brown color because of their high iron content.

Pillows of metamorphosed basalt come from lava that erupted underwater.
—U.S. Geological Survey photo, 1913 by B.L. Johnson

Tern Lake—Soldotna
58 mi./94 km.

A short distance from the junction of Tern Lake, the Sterling Highway skirts the edge of Kenai Lake. The lake occupies a glacially carved trough and is dammed by a terminal moraine. The highway then heads west along the canyon of Kenai River, passing outcrops of metamorphic rocks of the Chugach terrane and deposits of sand and gravel laid down by glacial streams. Cooper Creek was the site of the first discovery of gold in Alaska, in 1848 by Russian explorers. The Russian River has gold in it too, but there would be thousands of irate fishermen if any placer mines went near it. The last thirty-five miles into Soldotna are across the Kenai lowlands, which are almost completely covered by glacial deposits. They form a variety of glacial depositional landscapes and include till, outwash, and lakebeds.

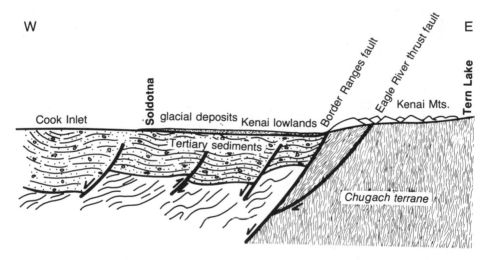

Cross section, Tern Lake–Soldotna

Till is the unsorted debris that glacial ice dumps as it melts. It includes all sizes of particles, from silt and clay through sand and gravel, to boulders. Glacially deposited landforms known as moraines are made of till. Outwash is stream-sorted material that has been transported and redeposited. It consists mostly of sand and

153

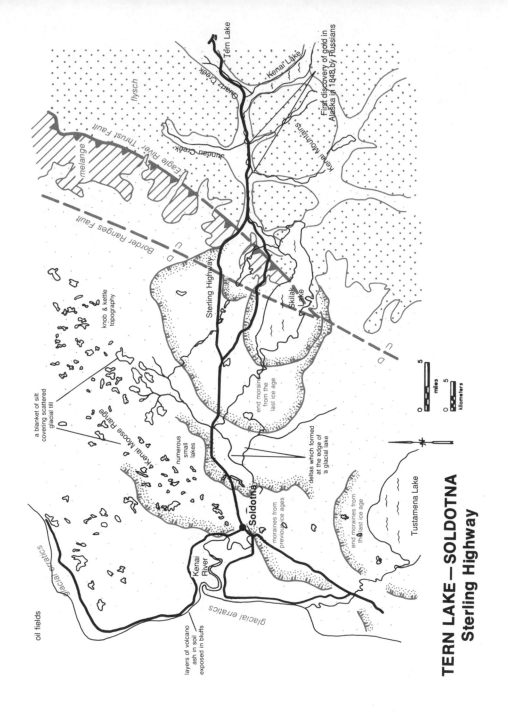

TERN LAKE—SOLDOTNA
Sterling Highway

Tern Lake

Quartz Creek

Kenai Lake

flysch

First discovery of gold in
Alaska in 1848 by Russians

Eagle River Thrust Fault

melange

Jundan Creek

Kenai Mountains

Border Ranges Fault

Sterling Highway

Skilak
Lake

U

D

knob & kettle
topography

D

U

a blanket of silt
covering scattered
glacial till

end moraine
from the
last ice age

Kenai Moose Range

numerous
small
lakes

5

miles

5

kilometers

deltas which formed
at the edge of
a glacial lake

Soldotna

moraines from
previous ice ages

end moraines from
the last ice age

Tustamena Lake

oil fields

glacial erratics

Kenai
River

glacial erratics

layers of volcano
ash in soil
exposed in bluffs

gravel with some cobbles. Usually it is layered and if the pebbles have a flat shape, they tend to be tilted upstream. Lakebeds are made of the fine glacial silt that makes water in glacial lakes milky. The bottom of any glacial lake receives a rain of silt and clay settling out of the water. Glacial lake beds drape over a large portion of the Kenai Moose range, creating poor drainage and swampy ground because they are relatively impermeable—bad for humans but good habitat for moose. The highway goes through lake beds along the Moose river near Sterling.

During Pleistocene time, all of Alaska and the Kenai Peninsula were periodically locked into ice ages; an ice cap covered the Kenai Mountains. All that is left of the ice today is the Harding Icefield. But during the Ice Ages the cap grew until it flowed down off the sides and, along with ice flowing from other ice caps, filled Cook Inlet to an elevation up to 4000 feet. The ice advanced and melted back at least seven times. Evidence of four major Pleistocene ice ages can be found on the Kenai Peninsula, evidence for three of them along this section of road.

Moraines of the last three ice ages remain on the Kenai lowlands between the mountains and Soldotna. A traveller coming down out of the mountains first encounters the moraines of the last Ice Age, followed by successively older moraines representing older, more extensive glaciations.

During the oldest, and greatest, Ice Age that we know about, ice covered the Alaska range, the Talkeetna, Chugach and Kenai mountains, and flowed into Cook Inlet, filling it to above 3000 feet in elevation. Ice from this and successive ice ages spilled out to the Pacific Ocean through the straits on either side of Kodiak Island. Evidence of this earliest Ice Age survives only high in the hills, such as the summit of Mt. Susitna west of Anchorage, above the ice from later ice ages.

The early Ice Age was followed by a long period of warm weather perhaps similar to what we have now, then came another Ice Age, whose glacial deposits are exposed on protected slopes in the Caribou Hills of the Kenai Peninsula. Another warm interglacial period followed.

The next Ice Age marked the last time Cook Inlet was completely filled with ice. Moraines from that Ice Age dot the lowlands around Cook Inlet and along the west coast of the Kenai Peninsula. The till is preserved because later ice advances did not go far enough to wipe it out. There are good exposures for about seven miles east of Soldotna.

During the following Ice Age, ice flowed down into, but did not completely fill, Cook Inlet. Huge piedmont glaciers much like those in modern Antarctica covered the mountain fronts and drained into Cook Inlet. Till from this glaciation is exposed near the road around Scout Lake.

The last major Ice Age left some well-formed end moraines east of Naptowne (now called Sterling). During the Ice Age, the Kenai ice cap flowed over, filling the Kenai River and Resurrection River valleys to elevations above 2000 feet. A glacier twelve miles wide stretched from the Kenai Mountains to Sterling. The moraine that dams Skilak Lake, and scattered till between the Russian River and Sterling, remains from this Ice Age.

The eruption of Augustine Volcano, 1986. —U.S. Geological Survey photo

Soldotna—Homer
77 mi./123 km.

South of Soldotna, the highway crosses old moraines before emerging on the coast near Clam Gulch. From there it follows bluffs that overlook Cook Inlet all the way to Homer. The bluffs are made of Tertiary sedimentary rocks that underlie the Kenai Peninsula and Cook Inlet. To the west across the water, volcanoes loom on the horizon.

Volcanoes

On a clear day, four active volcanoes are visible to the west across Cook Inlet. From north to south Mts. Spurr, Redoubt, Iliamna and St. Augustine dominate the western skyline. Past eruptions left layers of ash in the soil. The ash is light gray, fine-grained and feels gritty. In the bluffs of Captain Cook State Park at least half a dozen ash layers are visible. The volcanoes have all erupted recently and obviously could erupt again any time.

Mt. Spurr lies approximately 75 miles due west of Anchorage, north of the Chakachatna River valley and the Chigmit Mountains.

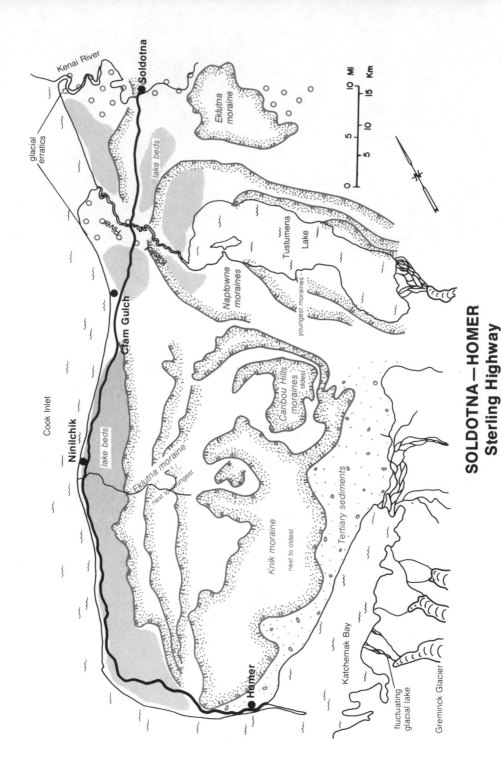

SOLDOTNA—HOMER
Sterling Highway

Kenai River

Soldotna

glacial erratics

Eklutna moraine

lake beds

River

Kasilof

Naptowne moraines

Tustumena Lake

youngest moraines

Clam Gulch

Cook Inlet

Caribou Hills moraines

oldest

Ninilchik

lake beds

Eklutna moraine

next to youngest

next to oldest

Knik moraine

Tertiary sediments

Katchemak Bay

Homer

fluctuating glacial lake

Greminck Glacier

Mi

Km

10

15

5

10

5

5

0

N

The knob rising from the high plateau is the summit of Spur. Some faint fumaroles are at the ice-covered summit, but recent volcanic activity has concentrated in a parasitic cone on the south flank of the mountain, visible in profile from the Kenai Peninsula. Steam rises from a turquoise lake inside the crater. In 1953 a spectacular eruption dusted Anchorage with as much as a quarter inch ash. Crater Peak on Mt. Spur began a new series of eruptions in August of 1992 that covered Anchorage in a fine dark brown ash, scoring car windshields as wipers attempted to remove the abrasive material. The Anchorage Airport was closed by the ashfall.

Iliamna and Redoubt are classic cone-shaped volcanoes, both active. In 1978, Iliamna ejected steam to a height of about 10,000 feet above its summit. Redoubt eruptions have been recorded in 1778, 1819, 1902, 1933, and 1965-68.

St. Augustine, the youngest and smallest of the four volcanoes, is the island visible on the distant skyline to the right of Kodiak Island and left of the Alaskan Peninsula. The volcano is probably less than 15,000 years old. It has a history of explosive eruptions, averaging 1 to 3 per century. A major eruption happened in 1976, when a glowing avalanche of greater than 600 degrees Centigrade roared down the slopes of the volcano. Winds blew the ash as far as 260 miles north, 600 miles to the east. More recently, in the spring of 1986, the volcano scattered a quarter inch of ash across Homer and shut down air travel in south-central Alaska. We should keep an eye on Mt. St. Augustine.

Blockade Lake refills itself after breaching its ice dam. The Kenai Peninsula is in the background. —U.S. Geological Survey by A. Post

The ebb and flow of the tides in Cook Inlet create extreme tidal currents.

Cook Inlet Tides

Cook Inlet has one of the greatest tidal fluxes in the world, almost thirty feet at the upper end. The surface waters are in continuous motion as the tides move in and out. After the high tide reaches Homer it will be almost four hours before the tide reaches Knik and Turnagain Arms at the north end of the inlet. The tides, influenced by the spin of the earth, flow in a counterclockwise rotation. The incoming tide flows north up the east side of the inlet and the outgoing tide flows south down the west side of the inlet. Consequently, mean sea level differs across the inlet.

Glacial Erratics

At Captain Cook State Park, the large boulders lying on the tidal flats are glacial erratics. Many are of rock types not typical of the Kenai: quartz diorite, meta-conglomerate, and schist. More large boulders are weathering out of till in the cliffs along the beach. The sea is gradually reclaiming the land, but the waves cannot move the big rocks.

Tertiary Sediments of Cook Inlet

Oil and gas production platforms are visible to the west in Cook Inlet, especially at night when the lights are on. Oil was first discovered on the Kenai Peninsula in 1957, at three miles depth near the headwaters of the Swanson River. Drilling later moved offshore, where much of the production is today. The wells penetrate Tertiary sedimentary rocks. A significant discovery of oil was made in western Cook Inlet in 1992. The Eocene Hemlock Conglomerate showed in tests that it contained up to 3,400 barrels per day. This was the first discovery since 1965 in the region.

The sand in the beach at Clam Gulch comes from Tertiary sediments in the bluffs overlooking the shore. Lignite beds on the bluffs are conspicuous. Fossil leaves and twigs are brittle but easy to find. The fossils are from plants in a mostly conifer forest that developed as the climate turned cooler at the end of Miocene time. The plants were buried along with the sediments in a river delta at the edge of the Cook Inlet basin between five and ten million years ago. These and other slightly older rocks of the Kenai formation are exposed in bluffs along the whole length of the Kenai Peninsula. Within a mile of the Homer Spit, there are 400,000,000 tons of low-grade coal in Tertiary sediments.

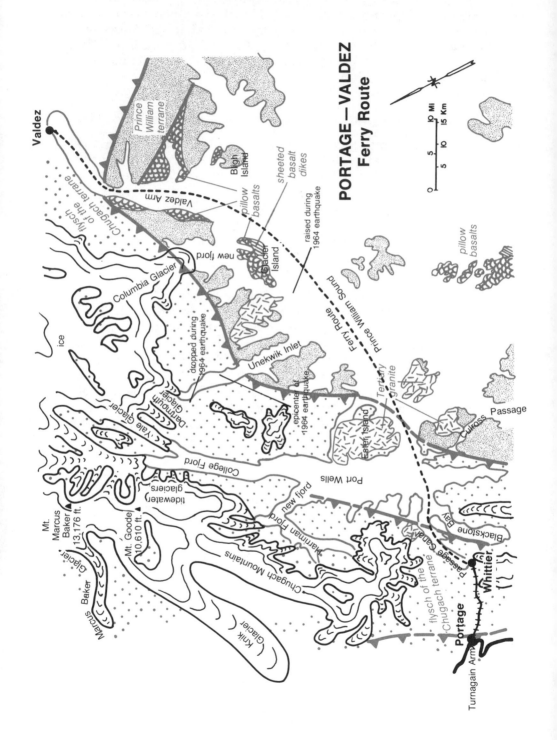

PORTAGE — VALDEZ
Ferry Route

Valdez

flysch of the
Chugach terrane

Chugach terrane

Prince
William
terrane

Bligh
Island

Valdez Arm

pillow
basalts

sheeted
basalt
dikes

new fiord

Glacier
Island

raised during
1964 earthquake

Columbia Glacier

ice

Ferry Route

Prince William Sound

Unekwik Inlet

dropped during
1964 earthquake

Dartmouth
Glacier

Yale Glacier

epicenter of
1964 earthquake

Esther Island

Tertiary
granite

pillow
basalts

Culross Passage

College Fiord

Port Wells

tidewater
glaciers

new fiord

Harriman Fiord

Mt.
Marcus
Baker
13,176 ft.

Mt. Goode
10,610 ft.

Chugach Mountains

Marcus Baker Glacier

Knik Glacier

Passage Canal

Blackstone Bay

Whittier

flysch of the
Chugach terrane

Portage

Turnagain Arm

0   5   10 MI

0   5   10   15 Km

This ship and its barnacles were left high and dry on Middleton Island after the 1964 quake raised the land by 30 feet. —U.S. Geological Survey photo by G. Plafker

PRINCE WILLIAM SOUND FERRY ROUTES

Portage—Valdez
100 mi./160 km.

Travel between Portage and Whittier is by train through a tunnel. Travel between Whittier and Valdez is by boat across Prince William Sound. The normal meaning of roadside is being stretched to accommodate the special demands of travel in Alaska.

Portage Glacier

Portage Glacier is barely an hour's drive from Anchorage and consequently is one of the most visited glaciers in Alaska. From the visitor's center, the view leads the eye across Portage Lake to the glacier beyond. Burns Glacier can be seen above the snout; the main body of Portage Glacier lies around the corner, to the right, where it comes down from Carpathian Peak.

163

Portage Glacier advanced as far as the end of the lake in 1914, but has since retreated almost 3 miles. The 595 foot deep lake is often filled with beautiful icebergs. The icebergs are unstable and should not be approached at close range, even when the surface of the lake is frozen. The glacier has been retreating as much as 50 feet per year; if that continues, there won't be any Portage Glacier visible from the new visitor's center by the year 2020.

Chugach and Prince William Terranes

Much of the bedrock in the railroad tunnels and in Prince William Sound is a dull, monotonous gray or gray-green but exposures are excellent because there are miles and miles of almost continuous outcrop in the wave-cut cliffs at the water's edge. The most common rocks are slightly metamorphosed volcanic rocks with their associated sedimentary deposits. The rocks of volcanic origin can be identified by their greenish color.

The rocks belong to two different terranes. North of Wells Passage and Valdez is the Chugach terrane, a collection of poorly sorted sediments and lavas from a rising volcanic island chain that formed about 130 million years ago somewhere in the Pacific Ocean. Far to the south lie Orca Group rocks of the Prince William terrane, a group of slightly younger rocks containing sediments, limestones, and basalts. Small granitic plutons are scattered here and there among the terranes, the product of melting of sediments above an active subduction zone.

Tidewater Glaciers

Prince William Sound contains the greatest collection of tidewater glaciers anywhere in Alaska; twenty active tidewater glaciers presently terminate at sea level. At a distance, the deep rumblings produced by ice crashing into the sea sound like distant cannon fire or thunder.

Many of the glaciers in this area were named by members of the 1899 Harriman Alaska Expedition. Glaciers in College Fjord and Columbia Glacier were named for colleges and universities associated with members of the expedition. In 1899 Barry Glacier, at the northwest end of Port Wells, almost completely blocked Barry Arm. There was just enough room to sail by the glacier, and in so doing, the expedition discoverd a previously unknown fjord, which they named Harriman Fjord. Surprise Glacier appeared to them as they rounded the corner into the fjord. Between then and 1914 Barry Glacier retreated more than four miles to near its present day position.

Most of the glaciers in Prince William Sound have been slowly retreating since the turn of the century, but there are some notable exceptions. Harvard, Harriman, and Meares glaciers have all been slowly advancing for the last half century. It is not clear why most glaciers should retreat while some advance.

Columbia Glacier, more than forty miles long and six miles wide at the terminus, is carefully watched because of its proximity to Valdez and the end of the Trans-Alaska Pipeline. Lately the glacier has been retreating from its terminal position where it rests in shallow water on an old moraine. As the glacier retreats into deeper water it is beginning to break apart, dumping large numbers of icebergs into the Valdez shipping channel, where supertankers carrying North Slope oil would be in jeopardy.

A great piedmont glacier occupied Prince William Sound during the last Ice Age. Many smaller glaciers flowed together to form a huge, bloated mass of ice that filled the sound and spilled out through Montague Strait, Hinchinbrook Entrance, and Orca Inlet. The ice was three to four thousand feet thick in much of the sound. The piedmont glacier probably bore some resemblance to the Malaspina and Bering glaciers that exist today along the Gulf of Alaska.

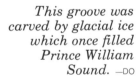

This groove was carved by glacial ice which once filled Prince William Sound. —DO

The Good Friday Earthquake

Even though Whittier was closer to the epicenter of the 1964 earthquake than Anchorage, it suffered less damage. Whittier, unlike Anchorage, is built on bedrock. Although the town was washed by great tsunami waves and received a terrible shaking, the residents of Whittier managed to ride out the earthquake in good shape because the ground didn't collapse beneath them as it did in some Alaskan towns.

Old time residents of Valdez divide time into either before or after the Good Friday Earthquake, which completely destroyed old Valdez and forced the construction of new Valdez. The old town, built on silt, was shaken violently for four minutes until cracks opened up in the ground and great blocks of land slid out to sea. The docks and buildings along the waterfront collapsed, and the damage was terrible everywhere in town. Finally, the town was hit by four tsunamis, earthquake-generated giant waves traveling through the water of Prince William Sound.

The wave nature of ground shaking was apparent to witnesses of the 1964 earthquake in Valdez. The ground appeared to move rhythmically up and down like swells on the surface of the ocean but more quickly. Buildings swayed so that sailors ashore felt like they were riding out a storm at sea. Trees swayed back and forth, and in the streets, fissures opened and closed repeatedly, spurting out water when they closed. Ground movement during an earthquake is caused by shock waves traveling through the solid earth away from the locus of fault movement.

Twenty five years after the March 27, 1964 Good Friday Earthquake, another disaster struck the town of Valdez. On March 24, 1989, the Exxon Valdez, a single-hulled 987-foot supertanker carrying oil from the Trans-Alaska Pipeline terminal in Valdez to west coasts ports in the lower 48 states, ran aground on Goose Island off of Bligh Reef in Prince William Sound, 25 miles south of Valdez. Eleven million gallons (250,000 barrels) of Alaska North Slope crude oil were spilled, ultimately impacting more than 700 miles of shoreline and the wildlife and people living along it. The spill damaged an area equivalent to the length of coastline stretching along the eastern seaboard of the United States from Cape Cod, Massachusetts, to Cape Hatteras, North Carolina. The Exxon Valdez oil spill is the largest yet to happen in North America, but will probably not be the last.

Earthquake waves striking
Whittier drove this plank
through a tire, 1964.

—U.S. Geological Survey photo

167

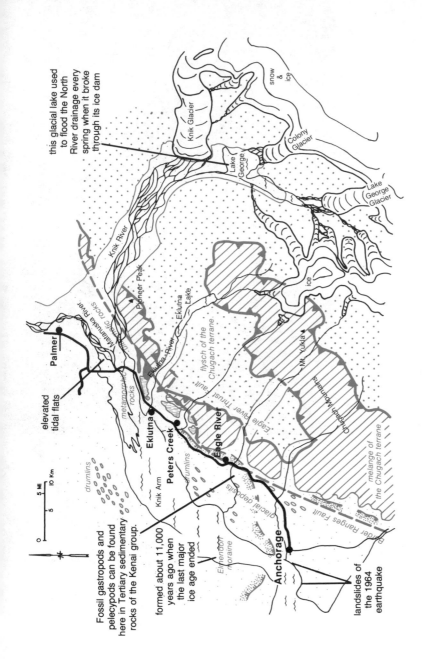

this glacial lake used to flood the North River drainage every spring when it broke through its ice dam

snow & ice

Knik Glacier

Colony Glacier

Lake George

Lake George Glacier

ice

Knik River

Matanuska River

Pioneer Peak

metamorphic rocks

Palmer

elevated tidal flats

Eklutna River

Eklutna Lake

flysch of the Chugach terrane

Eklutna

Peters Creek

Eagle River Thrust fault

Eagle River

Mt. Yukla

Chugach Mountains

Knik Arm

drumlins

drumlins

melange of the Chugach terrane

glacial deposits

Border Ranges Fault

Anchorage

Elmendorf moraine

Fossil gastropods and pelecypods can be found here in Tertiary sedimentary rocks of the Kenai group.

formed about 11,000 years ago when the last major ice age ended

landslides of the 1964 earthquake

0   5   5 MI
0   10 Km

ANCHORAGE—PALMER
Glenn Highway

THE GLENN HIGHWAY

Anchorage—Palmer
46 mi./64 km.

Glacial Landforms

The Elmendorf moraine ten miles north of Anchorage is the end moraine of the combined Matanuska and Knik Glaciers that advanced during the Naptowne Ice Age, beginning about 30,000 years ago. The ice maintained its position long enough to build the moraine, between 12,000 and 14,000 years ago, and then melted back about 11,000 years ago.

The Glenn Highway north of the Elmendorf moraine follows glacial deposits that mark the retreat of the Naptowne glacier. Lateral moraines of earlier, more extensve ice ages cling to the slopes of the Chugach above the reach of the Naptowne ice.

Mirror lake is a kettle pond formed by ground collapse when a large block of ice, half buried in till, melted and left a depression. Edmonds Lake occupies a lateral runoff channel formed during the Naptowne glaciation, and lower Fire Lake floods a glacially eroded trough.

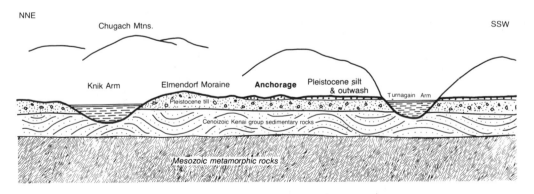

Cross section across glacial deposits of Anchorage lowlands.

Catastrophic Floods

Until 1966, the Knik River flooded every summer when Lake George overtopped its ice dam and cut a gorge through the ice against the valley wall. The river's flow was then augmented by up to thirty times normal as the glacial lake sought to empty itself. The event was so regular that the Alaska Railroad would put into its annual budget the projected cost of repairs. There has been no flooding since 1966 because the Knik Glacier has retreated to where it cannot dam the outflow of Lake George.

The Border Ranges Fault

The Border ranges fault runs north-south through Anchorage and along the front of the Chugach Mountains to Pioneer Peak, where it starts to bend eastward along the north side of the Chugach. The fault separates the Chugach terrane on the east from the Peninsular terrane on the west, and forms the eastern margin of Cook Inlet. The fault is considered active even though no large earthquake has yet been recorded along its length. The records don't go back very far. A trench across the fault near the town of Eagle River revealed soil profiles that were displaced, during a large earthquake about 300 years ago.

Bedrock exposures on the west side of the fault are part of the Peninsular terrane; there are good outcrops at Birchwood and at the turn-off of the old highway to Palmer. Rocks along the road are metamorphosed basalts and cherts from an early Jurassic volcanic chain. They contain early Jurassic igneous intrusions of intermediate composition; one of these can be seen in a rock quarry behind the village of Eklutna.

The old highway crosses the fault and goes by rock of the Chugach terrane near the Knik River. The rocks are wacke, a mixture of mud, sand, and gravel deposited in layers by turbid flows of sediment avalanching into deep water.

Elevated Tidal Flats

A few miles south of Palmer, the highway crosses tidal flats that are now exposed above sea level. There are several possible reasons for this. Elevation of the tidal flats may indicate an earlier, higher sea level, or it may be caused by tectonic forces, or by isostatic rising of the land. Isostatic rebound occurs after a great mass of ice that was holding the land down, is removed, permitting the crust to float up. Most likely a combination of these forces is at work.

Palmer—Glenallen
141 mi./226 km.

The Glenn Highway between Palmer and Glenallen follows the Matanuska River valley through road cuts in Mesozoic and Cenozoic sedimentary rocks, and passes deposits of the Matanuska glacier. The road winds its way between the Talkeetna Mountains on the north and the Chugach Mountains on the south. From Eureka Summit, the highway makes a beeline for Glenallen with the volcano Mt. Drum straight ahead on the skyline.

Sedimentary Rocks

The early Tertiary Chickaloon formation unconformably overlies the Matanuska formation and is well exposed in Anthracite Ridge and along the road east of Sutton. Sedimentary rocks of the Chickaloon formation are primarily continental in origin—mudstones, sandstones, and conglomerates. They were most likely deposited in alluvial fans along the mountain front at the edge of the basin. Coal seams in the Chickaloon formation appear in roadcuts, and were mined for many years on Anthracite Ridge. Fossil leaves are easy to find along Moose Creek, in roadcuts, and at the base of Anthracite Ridge. Ginkgo trees are common, and tell of a much warmer climate. The Chickaloon formation has been intruded by light-colored igneous dikes that are visible from the road.

King Mountain, on the south side of the highway, is granitic rock. The tabular pluton was intruded at depth into the hot rocks of the Chugach terrane in late Cretaceous or early Tertiary time.

The late Mesozoic Matanuska formation is well exposed on the road west of Sutton and where the old highway crosses the Matanuska River. The formation consists of shallow water marine sediments, primarily shales and mudstones that were deposited in deltas along the edge of the Cook Inlet sedimentary basin. Fossil ammonites and the large Cretaceous clam *Inoceramus* are common in the shale.

Gunsight Mountain, Anthracite Ridge, and Castle Mountain are all made of sedimentary rock that was deposited during Mesozoic and Cenozoic time. The Border Ranges fault parallels the road. The Chugach terrane to the south was shoved against the peninsular terrane to the north.

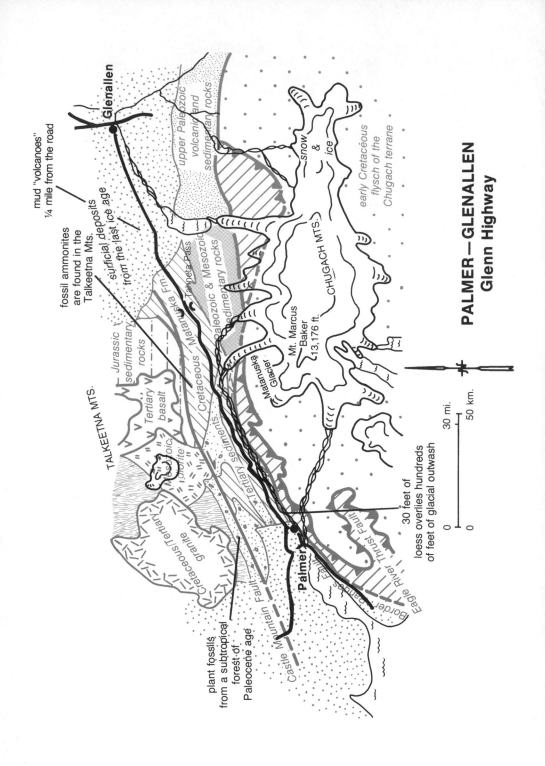

mud "volcanoes"
¼ mile from the road

Glenallen

upper Paleozoic volcanic and sedimentary rocks

fossil ammonites are found in the Talkeetna Mts.

surficial deposits from the last ice age

snow & ice

early Cretaceous flysch of the Chugach terrane

Jurassic sedimentary rocks

Tahneta Pass

Matanuska Fm.

Paleozoic & Mesozoic sedimentary rocks

TALKEETNA MTS.

Tertiary basalt

Mesozoic granite

Cretaceous

Tertiary sediments

CHUGACH MTS.

Mt. Marcus Baker 13,176 ft.

Matanuska Glacier

Cretaceous Tertiary granite

Castle Mountain Fault

Caribou Fault

Palmer

Border Ranges Fault

Eagle River Thrust Fault

30 feet of loess overlies hundreds of feet of glacial outwash

plant fossils from a subtropical forest of Paleocene age

0   30 mi.

0   50 km.

PALMER—GLENALLEN
Glenn Highway

Fossil ammonites of Jurassic age exist in the hills north of Eureka Lodge and in much of Castle Mountains. Ammonites, although their shells resembled those of snails, were cephalopods, tentacled predators related to octopus and squid that fed on shellfish. Ammonites could control their floatation and could move rapidly, if the occasion demanded it, by jet propulsion. They lived in shallow water on the continental shelf along the coast of Alaska, in seas that were warmer than Alaskan seas today.

Matanuska Glacier

You can drive right up to the Matanuska Glacier but it will cost you. This large, accessible glacier is reached by a privately owned access road and bridge. It is a good place to see a glacier dumping sediment. In summer, pools of mire guard the glacier and offer a shoe full of mud to the unwary. The Matanuska River is born brown with silt at the mouth of the glacier.

Look for glacial striations on boulders in the fresh till surrounding the parking lot. They are scratches left by glacial ice that rasped its way over hard surfaces, inscribing the marks of its passage. The lower end of the glacier is covered with rocks that were left standing there as the ice melted.

Glacial striations are made by ice crushing rocks against each other. Penny for scale. —DO

173

Eight thousand years ago, the glacier retreated up valley of its present location and then readvanced to its present position, where it has been, except for minor fluctuations, for the last seven thousand years. Stagnant ice on the north side has been there for the last 100-200 years. The stagnant ice has acquired an insulation layer of sediment on its surface and even supports a plant cover in places.

Between the Matanuska Glacier and Palmer the Glenn Highway follows the glacier trough eroded by the Matanuska Glacier during the ice ages. Moraines mark halting places in the retreat of the glacier up the valley. There is a good view from bluffs above the Matanuska River where the highway leaves the river before heading for Palmer. From this vantage the braided, silty Matanuska River approaches from the east. The Chugach Mountains are visible here on the south and the Talkeetna Mountains on the north. At this spot you are standing on thirty feet of loess, windblown silt picked up from the seasonally dry Matanuska and Knik rivers. Beneath the loess there is at least a hundred feet of glacial outwash. Sand dunes, and especially deposits of loess, are common to glaciated regions because all the barren glacial outwash and till is easily vulnerable to wind erosion. Winds pick up silt from the dry floodplains of the river and deposit it in the Palmer area.

The Tolsona mud volcanoes are actually warm springs whose saline waters have deposited silt in a broad, gentle mound. —DO

Mud Volcano

Tolsona Number One Mud volcano lies just a tenth of a mile north of the Glenn Highway, about 10 miles west of Glenallen, on top of the bluff east of Tolsona Creek. A primitive trail goes from the road through a black spruce swamp to the mud 'volcano,' actually a warm spring. It's about 25 feet high and as big around as two football fields. A ring of dead trees and patch of bare ground rim the summit. At the crest 30 to 40 vents bubble out methane gas a lukewarm water laced with sodium chloride and calcium chloride. The source of the metane gas is mysterious. Decomposing organic matter is a common source of methane, so it is possible that the gas comes from underlying coal beds.

A series of wildcat holes were drilled in the 1960s in a line between the Eureka Lodge and Gakona. The deepest holes penetrated Pleistocene, Tertiary, and Mesozoic sediments, ending in lower Jurassic volcanic rocks of the Talkeetna formation at a depth of almost 9000 feet. One drill hole across the highway from Tolsona Number One encountered warm water mixed with methane and tar in fine-grained quartz sandstone about 4500 feet below the surface. It seems plausible that the warm water and methane may be percolating to the surface and bubbling through lake silts at the surface to form the mud volcanoes.

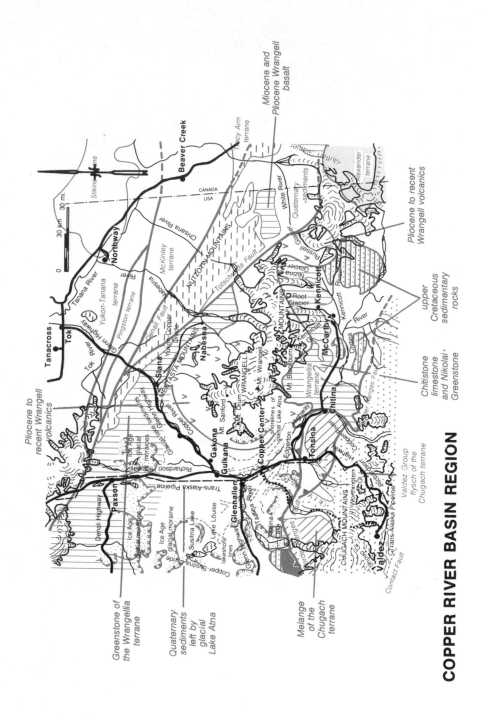

COPPER RIVER BASIN REGION

Greenstone of the Wrangellia terrane

Quaternary sediments left by glacial Lake Atna

Melange of the Chugach terrane

Pliocene to recent Wrangell volcanics

Miocene and Pliocene Wrangell basalt

Pliocene to recent Wrangell volcanics

upper Cretaceous sedimentary rocks

Chitistone limestone and Nikolai Greenstone

Tanacross

Tok

Beaver Creek

Northway

CANADA
USA

0   30 km
   30 mi

Stikine terrane

Tanana River

Tanana River

Tok River

Glenn Highway

Yukon-Tanana terrane

Pingston terrane

Denali Fault

McKinley terrane

Chisana River

Tetlin terrane

Totschunda Fault

White River

Nabesna River

Chistochina Glacier

Quaternary sediments

Russell Glacier

Skolai terrane

Nizina River

Alexander terrane

Malibu Glacier

NUTZOTIN MOUNTAINS

Yukon-Copper Divide

WRANGELL MOUNTAINS

Mt. Wrangell

Mt. Sanford

Mt. Blackburn

Root Glacier

Kennicott

Kennicott River

McCarthy

Kennicott

Chitina River

Chitina

Wrangellia terrane

Slana

MENTASTA MOUNTAINS

Nabesna

Copper River

Gakona

Gulkana

Copper Center

Mt. Drum

shorelines of glacial Lake Atna

Edgerton Highway

Tonsina

Copper River Highway

Richardson Highway

Copper River

Glennallen

Trans-Alaska Pipeline

Glenn Highway

Copper-Susitna Divide

Susitna Lake

Lake Louise

lakeshore lines

Tazlina River

Klutina River

Paxson

Ice Age glacial moraines

Glacial lake sediments

Denali Highway

Ice Age glacial moraine

Chugach Range

CHUGACH MOUNTAINS

Worthington

Valdez Group flysch of the Chugach terrane

Valdez

Trans-Alaska Pipeline

Contact Fault

River Ranges Fault

Elliptical cloud on the summit of Mt. Sanford (16,237 ft.), one of the Wrangell volcanoes. —National Park Service photo by M.W. Williams

Glenallen—Tok
135 mi./216 km.

Permafrost

Sections of the road between Glenallen and Tok have been attacked by frost heaving. Soils and subsoils in this region remain frozen both in winter and summer. Within the Copper River basin, southwest of Mentasta Pass, this permanently frozen ground, permafrost, exists everywhere except under large lakes and major streams.

The depth of frozen ground in this region ranges from 100 to 200 feet. You can reach it by digging from 1 to 2 feet beneath the surface in some muskegs or bogs that are well-insulated by thick Sphagnum moss. Permafrost lies 2 to 5 feet below the surface in fine-grained lake and glacial sediments, more than 6 to 10 feet below the surface in sandy and gravelly stream and glacial deposits. Much of the ice exists as solid masses, in lenses and layers, and occasionally as vertical wedges. The temperature of the permafrost is just below the freezing point. If the temperature is raised by removal of surface soils or vegetation, the underlying permafrost can melt, causing irregular subsidence of the ground surface. This creates unusual problems for construction projects and caused a great deal of "ice engineering" to go into building the Trans Alaska Pipeline.

Permafrost was kept out of the Copper River basin during Pleistocene time by the warming influence of a large glacial lake, but it has since reentered the basin and is especially widespread in areas covered by deposits of poorly-draining lake silt. The Glenn Highway's undulating surface in this region is due to the problems inherent in building roads on permanently frozen lake sediments.

Glacial Lake Atna

During one or more early Pleistocene glaciations, glaciers advanced from the Chugach, Wrangell, and Talkeetna mountains, and from the Alaska range into the Copper River basin where they covered the entire basin floor. Early in each phase of ice incursion from the Chugach Mountains, the Copper River was dammed by ice to form a large lake that varied in size as changing ice thicknesses controlled water outflow at the basin's major threshold areas, Mentasta Pass and the Copper-Susitna divide. Deposits of lake silt lie beneath much of the Glenn Highway. The vanished lake was named Atna after the Athabaskan name for the Copper River.

Gakona Junction

High bluffs along the Copper River, like these at Gakona, are Pleistocene glacial, volcanic, lake, and stream deposits that are now being re-exposed by the down-cutting of the modern Copper River. This erosion supplies a tremendous amount of suspended sediment to the river. Sediment loads average 789,700 long tons per day. The grade of the Copper River channel is ten times greater than that of the Yukon River, which drains the largely unglaciated country of interior Alaska.

Fossil pollen recovered from glacial sediments exposed in the river bluffs indicates that spruce forest similar to those that now grow in the basin thrived more than 40,000 years ago. Traces of pine pollen in these old sediments record pine forests growing downwind from Mount Wrangell at that time. Pine trees do not now grow in the Copper River basin, but they flourish to the northeast, in the Yukon, along the Gulf of Alaska, and in southeastern Alaska.

Fossil remains of the extinct woolly mammoth, the furry elephant that roamed Alaska during Pleistocene time have been found in a few places in the Copper River basin. These large mammals may have

A 1902 photo of Mt. Drum (12,010 ft.), from the field camp of one of the first U.S. Geological Survey mapping parties to study the northern and central Copper River basin. —U.S. Geological Survey photo by W.C. Mendenhall

entered the basin from the unglaciated areas to the north during interglacial intervals, and grazed along glacial outwash plains and river valleys.

Wrangell Volcanoes

Between Glenallen and Tok when the weather cooperates, there are excellent views of the Wrangell Mountains. Tanada Peak, Mount Sanford (16,237 feet), Mount Drum (12,010 feet), and Mount Blackburn (16,390 feet) lie to the southwest and south of the Nabesna Road Junction, about 65 miles from Tok; the Mentasta Mountains are to the east. The Copper River begins its journey to the Gulf of Alaska at the Copper Glacier on the northeastern slopes of Mt. Wrangell (14,163 feet). Mt. Wrangell, south of Mt. Sanford, is the youngest and most recently active volcano of the group.

Cross section through Wrangell Mountains.

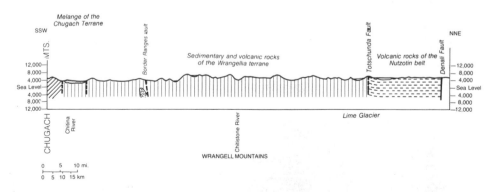

The Wrangell volcanoes began erupting during middle Miocene time, about 10 million years ago. Unlike the Aleutian volcanoes, which are the result of subduction of oceanic crust along the Aleutian trench, there is no subduction under the present continental margin south of the Wrangell Mountains, east of Prince William Sound. Wrangell volcanism is related to a small sliver of the Pacific plate sandwiched between the North American continent and the advancing Yakutat block, which began docking about 3 to 5 million years ago.

Mount Wrangell, the youngest and most active of the Wrangell volcanoes, was called Uk'eledi or "Smoking Top" by the Athabaskan residents of the region. It is one of the largest continental volcanoes in the world, a broad dome more than 14,000 feet high covered by a mantle of ice. The huge ice-filled basin or caldera at the summit covers nearly 12 square miles. Mount Wrangell has a volume at least six times larger than Mount Rainier in Washington state, and is the highest active volcano in Alaska. During the last 2 million years, it has erupted enough to cover about 3,500 square miles 300 feet deep. Mount Wrangell has not had a major eruption in historic time; the last major outburst happened between 2,000 and 10,000 years ago. The volcano has had several minor eruptions during this century.

Mount Wrangell is often called a shield volcano because of its dome shape, which closely resembles Mauna Loa on the island of Hawaii. That is unusual because volcanoes such as Wrangell that erupt viscous lava that piles up on the flanks of the mountain are normally conical, like Mt. Rainier. Geologists believe that Mt. Wrangell's rounded profile may be partly due to eruptions beneath glacial ice, which forces the lava to ooze laterally under the ice. Subglacial eruptions of this type are known to occur in Iceland, and are responsible for the flat-topped Togiak tuyas west of Dillingham, Alaska on Bristol Bay.

14,163 ft. Mt. Wrangell from the Government Trail south of Tonsina Bridge, Copper River basin, about 1902. —U.S. Geological Survey
photo by W.C. Mendenhall

Is Mount Wrangell likely to erupt in the near future? In 1899 a major earthquake shook Yakutat Bay, and knocked people down in Copper Center, south of Glenallen. Meanwhile, Mt. Wrangell blew off a dense cloud of ash. Large amounts of lava flowed down the northwestern slope for several miles, melting deep gorges in the snow and ice. Since the Alaska earthquake of 1964 the north crater of Mount Wrangell has been heating up, melting a huge chasm within the crater. The earthquake may have started the warm-up by re-aligning the plumbing system within the volcano. Active steam vents on the crater floor discharge heat energy at a rate equivalent to about 100 megawatts annually, enough to provide electrical power for a city of approximately 10,000 people.

Gas samples recently collected from the summit show high concentrations of sulphur dioxide, which may mean that magma is cose to the surface. Glaciers on the north side of the mountain have advanced, possibly lubricated at their bases by melt waters created by volanic heating. Additional hotspots have appeared around the mountain.

Road to Nabesna

In 1930 gold was discovered at Nabesna, southeast of the highway, by a prospector named Kal Wittum, who had wounded a bear and tracked it into a cave. He discovered quartz veins rich with gold and staked his claim, appropriately named the Bear Claims. This property operated until 1942 when the war closed it down. Mining has since resumed.

Mentasta Pass

Mentasta Pass is forty-six miles from Tok at an elevation of 2,434 feet. The Denali and Totschunda fault systems intersect in this area and have been extremely active in recent time. Right-lateral horizontal offsets in glacial deposits about 30,000 years old are as great as 274 feet near here.

Mentasta Pass was a key spillway for early drainage of glacial Lake Atna at the end of each Pleistocene glacial cycle. Lake levels were also controlled to the west at an elevation of 2,361 feet in the divide between the Copper and Susitna rivers. At lake levels below 2,361 feet, glacial lake Atna emptied to the south through the present Copper River valley near Chitina.

Twenty-one miles from Tok, the Glenn Highway crosses the Tok River. The size and breadth of the Tok River channel and alluvial fan suggest passage of a greater volume of water in the past than now. More than 10,000 years ago the Tok River acted as a spillway for an ice-dammed glacial lake Atna.

Tok lies at the junction of the Alaska and Glenn Highways, about 5 miles southwest of the confluence of the Tok and Tanana rivers. The Glenn Highway road segment passes out of the basin of the Copper River and crosses the Denali fault and Alaska range. Under the thick cover of Pleistocene sediments in the Copper River basin, are the Mesozoic rocks of the Wrangellia terrane and more volcanic rocks erupted from Mt. Wrangell during Pliocene and Pleistocene time.

Richardson Highway
Valdez—Paxson
186 mi./179 km.

Port Valdez

Like Skagway, Valdez served in 1898 as a supply center and base camp for Klondike miners seeking a route to the interior. The year-round ice-free harbor in the extreme northeastern fjord of Prince William Sound attracted many gold seekers bound for the Copper River and Yukon basins. In the spring and summer of 1898 and the early part of 1899, miners followed the Valdez trail up and over Valdez Glacier and Chugach Mountains, then northeast across Klutina Glacier to Klutina Lake in the Copper River basin. Miners who dawdled too long into the autumn in the Copper River basin soon faced starvation. Those who attempted to return to Valdez, recrossing the glacier in the winter, braved scurvy, deadly cold, and crevasses. The survivors who struggled back into Valdez convinced the local Army Commander Lt. Abercrombie that an alternative route was necessary and so in 1899 the military constructed a trail and telegraph line through Keystone Canyon along the Lowe River, southeast of Valdez. The trail crossed over Thompson Pass then led northward along the Tisina and Tiekel rivers, emerging west of the Copper River. That trail later became a wagon road and eventually the Richardson Highway.

On February 4, 1908 a strong earthquake shook the district and broke telegraph cables connecting Valdez with Seward and Seattle. A portion of the delta formed by the Lowe River and Valdez Glacier slumped, burying sections of cable lying in deep water on the fjord bottom about 11 miles from Valdez.

Valdez Group of the Chugach Terrane

Bedrock exposed in the fjord walls and canyons around Valdez is composed of several thousand feet of siltstone, sandstone, and conglomerate. These metamorphosed marine sections are part of the Valdez group, a portion of the Chugach terrane. These rocks correlate with the Sitka graywacke in southeast Alaska. The Port Valdez gold district, extending east about 24 miles, from Columbia to Valdez glaciers, was created during late Cretaceous and early Tertiary time as the Chugach terrane docked with North America. The collision created fractures, faults, and shear zones that later filled with gold-bearing quartz veins.

Copper River Basin

During Pleistocene time, expanding icesheets and glaciers in the Chugach Mountains dammed the Copper River, creating huge Glacial Lake Atna in the Copper River basin. Evidence for this glacial lake exists throughout the basin. Where the Richardson Highway nears Tonsina just north of Pump Station 12 on the Trans Alaska Pipeline, a sequence of Ice Age glacial lake sediments is exposed in the road cuts. Opposite Tonsina Lodge, a thick sequence of varved, or thinly layered, fine silts was deposited in a long-lived, deep lake that received large volumes of glacial sediment and iceberg-rafted pebbles and cobbles.

Border Ranges Fault

Westward from Tonsina across Tazlina Lake and on west to Matanuska Glacier, is a belt of mafic rocks of Jurassic age at least 6 miles wide and 72 miles long. The rocks mark the position of the Border Ranges fault, northern limit of the Chugach terrane. They possess a very strong magnetic signature. Magnetic studies of the area made from aerial surveys enable geologists to trace the fault beneath the thick cover of Ice Age glacial and lake sediments in the southern Copper River basin.

Between the Border Ranges fault and the Tazlina fault, near the south end of Tazlina Lake, is a melange of scrambled rocks known as the McHugh Complex. These oceanic crustal rocks include gabbros and greenstones, with cherts mudstone, and graywacke sandstone,

184

stratigraphically on top. As the Chugach terrane docked with North America, fragments of the continental margin were scraped off and incorporated into the McHugh Complex, adding further to this hodgepodge of rock types. The rocks were slightly metamorphosed to produce the low-grade minerals prehnite and pumpellyite. Fossil radioloria from chert samples within the McHugh Complex date its arrival during late Cretaceous time.

The bedrock south of the Tazlina fault is part of the Valdez group, mainly slightly metamorphosed shale and sandstone. To the south along the crest of the Chugach Mountains, metamorphosed sandstone predominates. The Valdez group was jammed beneath the McHugh rocks during a later stage of Chugach terrane docking, and metamorphosed to develop greenschist minerals.

A lahar or volcanic mudflow near the confluence of the Tonsina and Copper rivers, 1979. —U.S. Geological Survey photo by K.M. Johnson

Edgerton Highway 10
Southeast to Chitina and McCarthy

Glacial, volcanic, and lake sediments deposited during the ice ages thickly cover the central part of the Copper River basin. A brightly colored mosaic of transported volcanic rock appears in road cuts along the Edgerton Highway, in the Kotsina Delta area opposite Chitina, and along the road to McCarthy. The roadways slice through a giant volcanic mudflow that probably originated near Mt. Wrangell and moved west and south down the Chetaslina River and into the Copper River. The mudflow consists of a rainbow colored matrix of soft clay

Andesite flow with columnar base near Mt. Wrangell along the Chetaslina River 1979.

—U.S. Geological survey

photo by K.M. Johnson

material enclosing volcanic rocks that range in size from pebbles to boulders the size of a house. The mudflow must have been enormous, it reaches at least 42 miles south of Mt. Wrangell. It flowed into Glacial Lake Atna while the Copper River was still dammed by ice, no more than 200,000 years ago. Lake silt was deposited on top of the mudflow at Willow Creek, near the community of Kenny Lake on the Edgerton Highway.

McCarthy and the Copper of the Copper River Basin

From the town of Chitina, a bridge across the Copper River connects the Edgerton Highway with a dirt road leading to the settlement of McCarthy, famous for its copper mines. The first miners in this region were the Indians of the White River, who traded native copper with the Indians of Prince William Sound and with the Chilkat Indians near Haines. The Chilkats controlled access to the interior from the northern panhandle and effectively kept Lord Baranof's Russian traders from trading inland. The most important copper discovery in Alaska was made on the east side of Kennicott Glacier near McCarthy. In 1885 Lt. Allen, a military explorer, was introduced to Chief Nikolai of Taral, and under his guidance explored the Copper and Chitina rivers. The chief showed Allen his personal copper vein as well as some of its by-products including bullets made of copper-silver alloy, huge nuggets of native copper, and copper cooking utensils.

The mill at Kennicott about 1914. —U.S. Geological Survey photo by A.H. Brooks

In 1900 two Valdez prospectors discovered the Bonanza Mine in the same area; later the Chitina Mining and Exploration Company was formed. The world's richest copper deposit was 120 miles from the coast and accessible only on foot over some of the most rugged terrain in Alaska. To build the Copper River and Northwestern Railroad from Cordova up the Copper River to Kennicott, Guggenheim Transportation and Shipping hired the railroad engineer who built the White Pass and Yukon Line between Skagway and Whitehorse. The mine operated from 1911 to 1938 and produced more than $200,000,000 of copper—at today's prices that would be more than $1,000,000,000.

The copper ore occurs as fillings in the gas-escape holes of the Nikolai basalt of the Wrangellia terrane. Later some of the copper migrated upsection into the overlying Chitistone limestone. The limestone formed in a tropical marine environment under conditions resembling those in the Persian Gulf today.

North on the Richardson Highway from Copper Center

Several thousand goldseekers crossed Valdez Glacier and the Chugach Mountains into the Copper River basin between 1898 and 1899, and emerged at Klutina Lake, wintering between there and Copper Center. The installation of a telegraph in 1901 made Copper Center the major supply town in the Nelchina-Susitna region.

History of a Riverbank

Just north of the Tazlina River near its junction with the Copper River at Glenallen, a sharp bend in the Copper River exposes about 260 feet of Pleistocene and recent sediments. This steep river section records a dynamic history of an Ice Age river, the passing of a volcanic mudflow, the growth of a glacial lake with many icebergs, the arrival of Chugach Moutain glacial ice sheets, some volcanic activity, a second invasion of glacial ice, and the final period of glacial lake filling and draining that allowed the Copper River to flow again. Geologists can read a lot of history in a good cross section. A look across the river here provides excellent views of the Wrangell Mountains and some mud volcanoes along their lower slopes.

North of the Glenallen turnoff and Gakona Junction, the Richardson Highway follows the Gulkana River to Paxson Lake. The discontinuous permafrost of this region, in combination with widespread glacial lake silt, makes poor foundation material for both the highway and the Trans Alaska Pipeline, which passes close by. Where possible, the pipeline was elevated over permafrost areas but near Hogan hill at mile 155.2, the highway and pipeline cross an important caribou migration route and the pipeline was buried. To prevent the permafrost from thawing, the pipeline was heavily insulated and refrigerated with circulating brine, pumped from a plant adjacent to the highway.

Near Meier Roadhouse, exposed granitic bedrock is much broken by frost heaving. This area was north and above the level of Glacial Lake Atna and south of the ice moving out of the Alaska range. Farther north on the Richardson Highway near the eastern shore of Meier Lake, a roadcut reveals fractured greenstone that overlies the Wrangellia terrane. This greenstone also outcrops in roadcuts along the eastern shore of Paxson Lake. Between Meier and Paxson Lakes, the highway is underlain by glacial moraines that were left by the last major advance of ice from the Alaska Range, between about 30,000 and 10,000 years ago.

The surging Black Rapids Glacier in September 1937, about 2 miles south of the Rapids Roadhouse. —U.S. Geological Survey photo by F.H. Moffit

Paxson—Delta Junction
80 mi./128 km.

The Richardson Highway north of Paxson travels parallel to the Pipeline, follows the Delta River north through an ice-carved valley in the Alaska range, continues down the other side, goes around a big moraine, and then meets the Delta River again at Delta Junction. Several large glaciers are visible in good weather where the road crosses the highest part of the range.

Rainbow Mountain

The reds and greens in Rainbow Mountain are volcanic rocks and the yellows and pastels are siltstone and sandstone. Hot andesite and rhyolite lavas were squeezed in between sedimentary rock layers. We know from age dates that the lavas erupted in Pennsylvanian and Mississippian time. Their present location along the Denali Fault means that the rocks have been dragged a long distance.

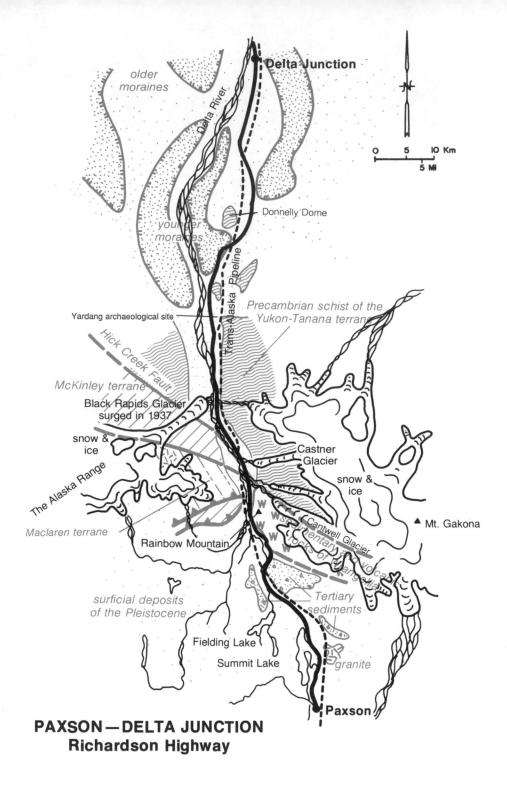

older moraines

Delta Junction

Delta River

0 5 10 Km

5 Mi

younger moraines

Donnelly Dome

Trans-Alaska Pipeline

Precambrian schist of the Yukon-Tanana terrane

Yardang archaeological site

Hick Creek Fault

McKinley terrane

Black Rapids Glacier surged in 1937

snow & ice

Castner Glacier

snow & ice

The Alaska Range

Mt. Gakona

Maclaren terrane

Cantwell Glacier

Rainbow Mountain

sedimentary and volcanic rocks of the Wrangellia

surficial deposits of the Pleistocene

Tertiary sediments

Fielding Lake

Summit Lake

granite

Paxson

PAXSON — DELTA JUNCTION
Richardson Highway

Prehistoric Inhabitants

The pass through the moutains has had human traffic for the last two thousand years. We know this because stone and bone artifacts have been found near the mouth of a creek where it enters the vegetated floodplain of the Delta River. People probably camped there during the caribou migration and ambushed the animals as they came through the pass. Their tools included flint scrapers and knives.

Surging Glacier

Most glaciers flow at the rate of a foot or two per day, but once in a while, for reasons unknown, a glacier will 'surge.' A surging glacier advances at a hundred times its normal rate and then, overextended, dies in its tracks. In 1927 the Black Rapids Glacier surged, and attracted worldwide attention as it flowed forward at up to 200 feet per day, advancing near the highway and the Black Rapids Roadhouse. During the surge, a radio announcer waited at the roadhouse to broadcast the details of its destruction. In dramatic fashion, the glacier stopped just short of the highway and roadhouse.

Denali Fault

Readers who have been to Denali Park may know of the great fault there that cuts the park in half and allows the high mountains of the Alaska range to rise. The same fault, the Denali fault, crosses the Richardson Highway north of Rainbow Mountain and then goes up the Cantwell Glacier, and continues on across to the southeast Canadian border.

Terranes

The best exposures of bedrock on this road segment are where the road goes though the heart of the Alaska range north of Rainbow Mountain. On the north side of the Denali fault you are on the Yukon-Tanana terrane, and on the south side, on Wrangellia. To the west there is a wedge of McKinley terrane between the two, seen also on the road between Cantwell and Fairbanks.

The McKinley terrane probably once resembled the modern Alaskan seamounts which are now approaching Alaska on the Pacific plate. The terrane is made of rocks from the floor of the ocean and from volcanic sediments deposited on the ocean floor. The eroded remnants of the volcano and the sediments derived from it were plastered against the continent in late Mesozoic time, along with pieces of the ocean floor.

The Wrangellia terrane is an old volcanic island chain that was born near the equator—and slowly subsided while coral reefs and lagoons formed around it. Today in this terrane we find volcanic breccias, flows, and eroded volcanic material, overlain by limestones and shales.

The Yukon-Tanana terrane is the name given to a very broad area containing rocks that have been strongly metamorphosed several times and now bear not the slightest resemblance to their former selves. Exposed here on the north side of the Denali fault, the original sedimentary rocks, perhaps one billion years old, have been changed to schist, gneiss, and quartzite. During metamorphism, sedimentary rocks containing clays and feldspars are changed to schist and gneiss, and sandstones changed to quartzite.

GEORGE PARKS HIGHWAY

Palmer—Cantwell
170 mi./274 km.

The highway between Palmer and Talkeetna passes through forest most of the way, so the geologic scenery is limited to what may be seen in road cuts. West of Palmer on the way to Willow, the route crosses glacial drift. North of Willow to Talkeetna, the road passes by glacial outwash. A widespread layer of loess blankets the glacial deposits. The highway skirts Cook Inlet basin, a large structural and physiographic basin that is mostly covered by seawater. The basin has been receiving sediments from the surrounding mountains for the last 100 million years; the total accumulation of sediment may amount to as much as 30,000 feet.

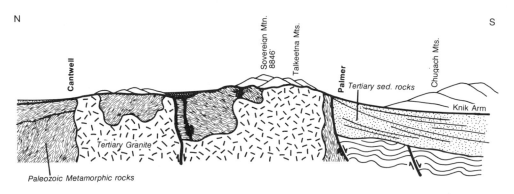

Cross section between Cantwell and Palmer.

Glacial Deposits

Between Palmer and Willow, glacial drift was deposited from a lobe of ice retreating up the Matanuska Valley during the last Ice Age. Numerous small lakes dot the area, kettle ponds formed by ice blocks melting out from within the glacial deposits. About five miles west of Wasilla there is a large group of drumlins. These elongate mounds of

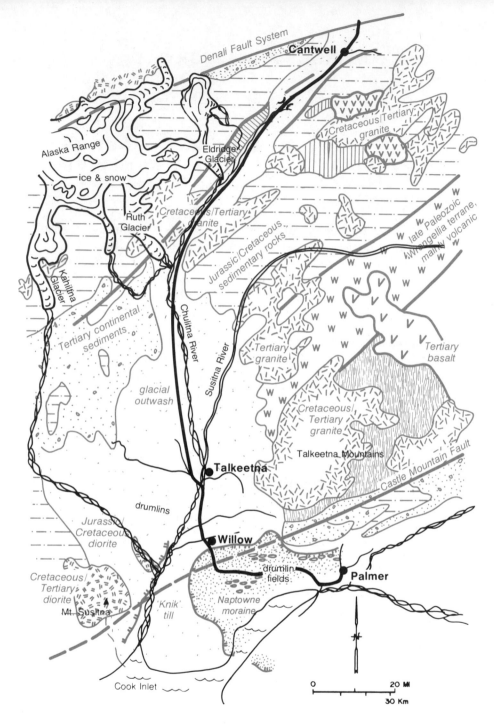

Denali Fault System

Cantwell

Alaska Range

Cretaceous/Tertiary granite

Eldridge Glacier

ice & snow

Cretaceous/Tertiary granite

Ruth Glacier

Jurassic/Cretaceous sedimentary rocks

late Paleozoic Wrangellia terrane, mafic volcanic

Kahiltna Glacier

Tertiary continental sediments

Chulitna River

Susitna River

Tertiary granite

Tertiary basalt

glacial outwash

Cretaceous/ Tertiary granite

Talkeetna Mountains

Castle Mountain Fault

Talkeetna

drumlins

Jurassic/ Cretaceous diorite

Willow

Cretaceous/ Tertiary diorite

drumlin fields

Palmer

Mt. Susitna

'Knik' till

Naptowne moraine

0   20 MI
30 Km

Cook Inlet

PALMER—CANTWELL

till often appear in swarms. It is not entirely clear how drumlins are created, but they occur where glacial deposits are overridden by a broad glacial sheet.

Willow stands on the edge of a huge glacial outwash plain that extends south from the Alaska range. Rivers that flow into the Susitna River from the west are born of great glaciers in the Alaska range and carry a temendous load of sediments to the Susitna River. If you place your ear close to the river you can hear sand scraping along the bottom.

Between Palmer and Cantwell the Parks Highway crosses the southern edge of the Alaska range and Denali Park and follows the Chulitna River upstream. Most of the great glaciers in this section of the Alaska range are on the south side because storms generally approach from the Pacific and dump their moisture as they rise over the mountains. The highway approaches quite close to the Ruth and Eldridge glaciers which are retreating. The lower part of the Eldridge glacier is so stagnant that in places its surface is covered with a thick growth of vegetation.

During the ice ages, glaciers flowed down all the way into the Chulitna drainage and combined there into a huge glacier that flowed south from Broad Pass, probably at least as far as Talkeetna. Numerous eskers, drumlins, and other glacial landforms attest to the movement of the massive ice lobe.

The part of the Alaska range visible from the highway is made of sedimentary rocks—mostly shales and mudstones deposited during Jurassic and Cretaceous time when an earlier range of mountains was forced up by continental collision, and then eroded. It seems logical to draw an analogy between these old sediments and younger ones now accumulating in Cook Inlet and Bristol Bay. Perhaps some day these modern sediments will be forced up into a great mountain range. As the sedimentary rocks in the Alaska range were pushed up, granitic magma worked its way up into the sedimentary rocks to form scattered intrusions such as those on the summit of Denali (Mt. McKinley) and in the Ruth Gorge.

South of Cantwell the Parks Highway goes through long narrow hills that run parallel to each other with long, skinny lakes between them. The road follows the crest of one hill for over two miles. These drumlins were shaped by ice flowing parallel to their length.

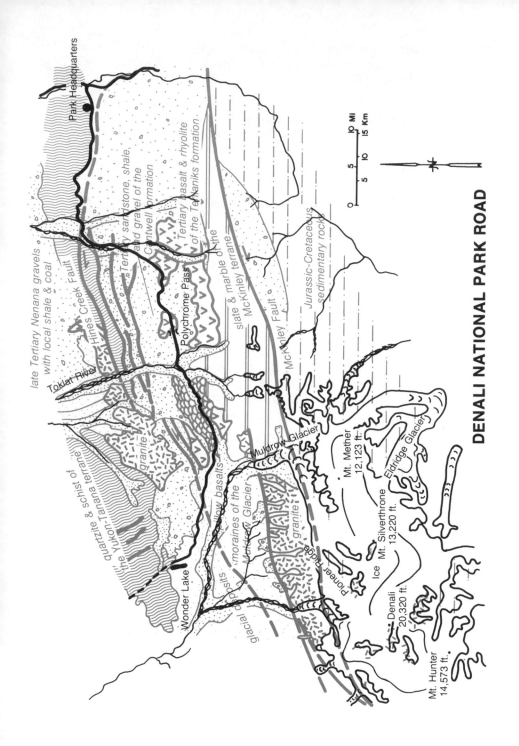

DENALI NATIONAL PARK ROAD

Park Headquarters

late Tertiary Nenana gravels
with local shale & coal

Hines Creek Fault

Toklat River

Tertiary sandstone, shale,
and gravel of the
Cantwell formation

Polychrome Pass

Tertiary basalt & rhyolite
of the Teklaniks formation

slate & marble of the
McKinley terrane

Jurassic-Cretaceous
sedimentary rocks

quartzite & schist of
the Yukon-Tanana terrane

granite

McKinley Fault

Muldrow Glacier

Mt. Mather
12,123 ft.

Wonder Lake

glacial deposits
moraines of the
McKinley Glacier

valley basalts

granite

Pioneer Ridge

Ice

Mt. Silverthrone
13,220 ft.

Eldridge Glacier

Denali
20,320 ft.

Mt. Hunter
14,573 ft.

5 MI
5 Km

0

5

5

10

10

10 MI
15 Km

196

DENALI NATIONAL PARK ROAD

Denali Fault System

The Hines Creek fault is an offshoot of the Denali fault system, a major continental translocation that can be traced clear across the state of Alaska. The fault passes near park headquarters and runs from east to west, along the line of the road as far west as the Teklanika River, where the road turns south. The road and fault rejoin near the end of Muldrow Glacier, to the west. North of the fault are schists and other metamorphic rocks of the Yukon-Tanana terrane, rocks as much as a billion years old. South of the fault is a band of younger metamorphic rocks, part of the McKinley terrane.

Tertiary Sediments

Along the eastern end of the park road are conglomerate, sandstone, and shale of the Cantwell formation. These sediments eroded from mountains that rose as terranes to the south docked with ancient Alaska. They accumulated in basins as nearby blocks of the crust were uplifted and eroded. Colorful lavas of the Teklanika formation erupted in the final stages of the uplift.

Polychrome Pass

Colorful rocks of Polychrome Pass are rhyolitic lavas that erupted about 60 million years ago. A short distance down the road is granitic rock, also about 60 million years old. Granite and rhyolite are similar in composition, but rhyolite is volcanic and forms at the surface; granite is plutonic and forms deep in the earth. It is likely that the granite and rhyolite both formed from magma that melted when the McKinley terrane docked with old North America.

Alaska Range

The Alaska range has a granitic core. Mounts Hunter, Foraker, the Moose's Tooth, and Denali (Mt. McKinley) are all made of granite intruded during Mesozoic time. Most of the other big peaks in the park are made of heavily metamorphosed sedimentary rocks. Many of the lesser peaks on the south side of the range are sedimentary rock, mainly shale and mudstone, deposited during Jurassic and Cretaceous time, then very slightly metamorphosed.

It is not exactly clear why Denali (Mt. McKinley) is so big. Its isolated mass rises 15,000 feet above the surrounding countryside. Denali's great height must be related to its location at a bend in the Denali fault system. The crust probably thickened there as one crustal block shoved against another. Some geologists have postulated the existence of a buoyant block of continental crust dragged under the continent by subduction that now is pushing up from below.

Muldrow Glacier

The Muldrow Glacier is visible on the south side of the road on the way to Wonder Lake. It is dying. Stagnant parts of the glacier have not moved in so long that brush and even trees have started to grow on its surface. Alaska is a real jungle in the summer, and if given a chance, the brush will grow even on the thin layer of debris covering stagnant ice. Landforms documenting the retreat of a larger Muldrow Glacier dot the countryside on the way to Wonder Lake. Today the little McKinley River winds through an oversized valley once occupied by the Ice Age Muldrow Glacier. A lateral moraine is visible to the south across the river; the road goes over its twin.

Looking east across the main Teklanika Glacier. By 1993 the Teklanika Glacier had receded back up the valley and almost out of view. —U.S. Geological Survey by S.R. Capps

Cantwell—Fairbanks
148 mi./238 km.

From Cantwell, the Parks Highway follows the Nenana River through the Alaska Range to Nenana. Where the road passes through the Alaska range, roadcuts reveal slices of crust that have been displaced along the Denali fault system. Farther north, Tertiary sediments contain coal. The road between Nenana and Fairbanks continues through the Tanana river lowlands before looping northeast along the north side of the Tanana River through rounded hills of schist.

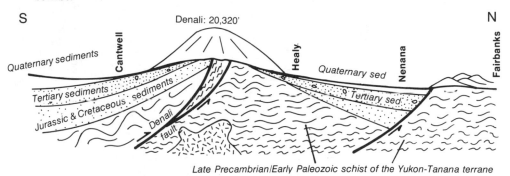

Cross section through the Alaska Range between Cantwell and Fairbanks. The Denali fault system places older schist of the Yukon-Tanana terrane alongside younger sedimentary rocks.

Old Terranes

The McKinley terrane now occupies the downdropped space between two large parallel faults. The Hines Creek fault forms its northern boundary. It is an offshoot of the Denali fault system that passes very close to park headquarters. The McKinley fault forms the southern boundary; it too is part of the Denali fault system, and crosses the road about five miles north of Cantwell.

On a geologic map, the McKinley terrane resembles a roll of dough that was stretched until it snapped into three segments. The segment on this road thins to the east until it pinches out near the Black Rapids Glacier north of Paxson. To the west, a segment underlies the Denali Park Road. A third segment lies far to the east, near the

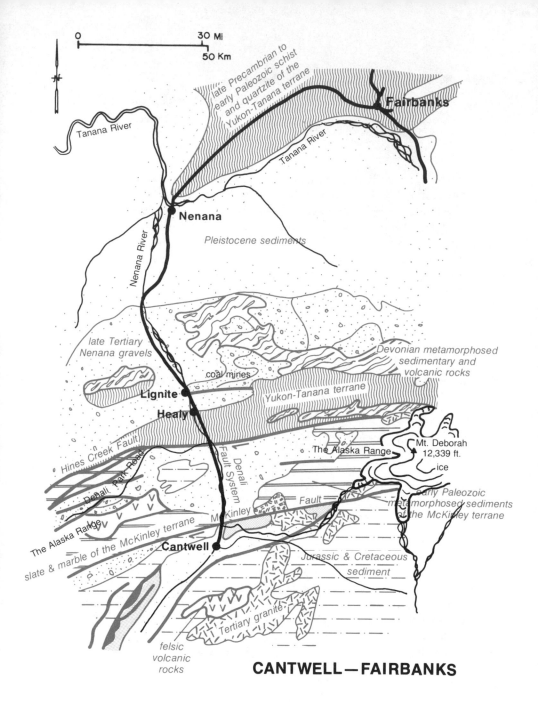

late Precambrian to early Paleozoic schist and quartzite of the Yukon-Tanana terrane

Tanana River

Tanana River

Fairbanks

Nenana

Nenana River

Pleistocene sediments

late Tertiary Nenana gravels

coal mines

Devonian metamorphosed sedimentary and volcanic rocks

Yukon-Tanana terrane

Lignite

Healy

Hines Creek Fault

The Alaska Range

Mt. Deborah 12,339 ft.

ice

Denali, Park Road

Denali Fault System

early Paleozoic metamorphosed sediments of the McKinley terrane

Fault

The Alaska Range

McKinley terrane

McKinley

Cantwell

slate & marble of the McKinley terrane

Jurassic & Cretaceous sediment

Tertiary granite

felsic volcanic rocks

CANTWELL—FAIRBANKS

Tertiary sandstone and conglomerate layers are interbedded with coal on lower Healy Creek, 1914. —U.S Geological Survey photo by S.R. Capps

Yukon border. This stretched out terrane is a collection of late Paleozoic impure sandstone, and early Mesozoic chert, pillow basalt, and gabbro. On top is deformed, impure sandstone and chert. Rocks of the McKinley terrane tell the story of a piece of the earth's crust, mostly oceanic but with some andesitic volcanoes that had drifted against the continent by late Mississippian time, then was squeezed and pulled like a very large piece of dough.

The Cantwell formation that overlies the McKinley terrane is composed of conglomerate, sandstone, shale, argillite, and coal. These sediments were deposited in late Cretaceous time, then later intruded by igneous dikes, sills, and laccoliths ranging from dark-colored mafic rocks to light-colored felsic rocks. Tertiary conglomerate and sandstone, and coal-bearing sediments overlie the Cantwell formation; there are coal mines in the hills above Healy and Lignite.

About a mile north of the park entrance, the road crosses the Hines Creek fault; for a couple of miles beyond, the roadcuts contain schist and phyllite, with quartz veins in interesting crenulations and folds. The silvery mineral is muscovite mica. These metamorhpic rocks are part of the Yukon-Tanana terrane, described in several other sections of this book.

Ice-Age Deposits

North of the park headquarters, the highway crosses a broad alluvial fan complex covered with glacial outwash. At milepost 303 south of Nenana there are vegetated sand dunes. A little farther north a good view opens to the west of lowlands containing many small lakes. The lakes are in scattered glacial till that has been mantled with silt.

Hills north of Fairbanks are capped with wind-blown loess and hold large masses of ground ice. Permafrost in fine-grained sediment like loess is generally more troublesome than permafrost in other sediments because water is drawn into the pore spaces by capillary action and ice can accumulate. Building on permafrost is fraught with geologic hazards, as residents of Fairbanks have discovered over the years. 'Thermokarst' features develop when permafrost is thawed, shrinks, and the ground subsides, usually unevenly. Thawing of permafrost can be induced by clearing vegetation, which has insulated the ground, or by heat escaping from buildings placed directly on the ground.

On the north side of the University of Alaska campus, almost all of Farmer's Loop Road lies on ice-rich permafrost. Road maintenance problems due to thawing of permafrost and differential settling have been so persistent that portions of the road and other nearby roads have had to be relocated.

Gold Placers

There are placer mines along Ester Creek west of Fairbanks; some use a dozer and dragline next to the highway. A thick silt overburden must be removed to get down to the frozen, gold-bearing gravels. There is little glamour in gold mining and a lot of hard work. Dredges started operating in Ester Creek in the 1920s and continued until the 1960s. The coarse gold-bearing gravel is made of pieces of phyllite, slate, schist, and gneiss.

Prehistory

In the 1930s stone age artifacts were found on the edge of the hill that the University of Alaska now occupies. The Campus site was the first in North America to produce artifacts that resembled ones found in Asia. Flint blades and cores resemble, in design and workmanship, their counterparts in the Gobi Desert - and lend support to the idea that humans first migrated to North America from Asia. During the Ice Age, Fairbanks and most of the Interior were free of ice, so humans and other animals could live there.

A great many vertebrate fossils have been found around here in frozen silt, enough for us to know that the area was rich with game. Bison are the most abundant large mammal remains around Fairbanks, followed by the mammoth and the horse. Also present were species of large cats, musk oxen, wolves, elk, sheep, moose, and mastodons. Radiocarbon dates indicate ages between 12,000 and 30,000 years. Some species that had survived through several ice ages and interglacial ages, died out at the end of the last Ice Age.

Gold nugget from Lucky Gulch near Valdez Creek, 1910. —U.S. Geological Survey photo by F.H. Moffit

DENALI HIGHWAY

Cantwell—Paxson
136 mi./218 km.

In summertime the Denali Highway arcs across the southern slope of the Alaska range. The western half of the route overlies Jurassic to Cretaceous sedimentary rocks, the eastern half early Tertiary basalt. The entire area was covered by ice during the last Ice Age, so there are plenty of ice sculpted landscapes.

Sand Dunes

Watch for sand dunes the first few miles out of Cantwell. The climate doesn't have to be hot and dry for dunes to develop, as long as weather is dry once in a while and there is a good supply of sand. There is a lot of loose sand in the braided stream channels of glacial

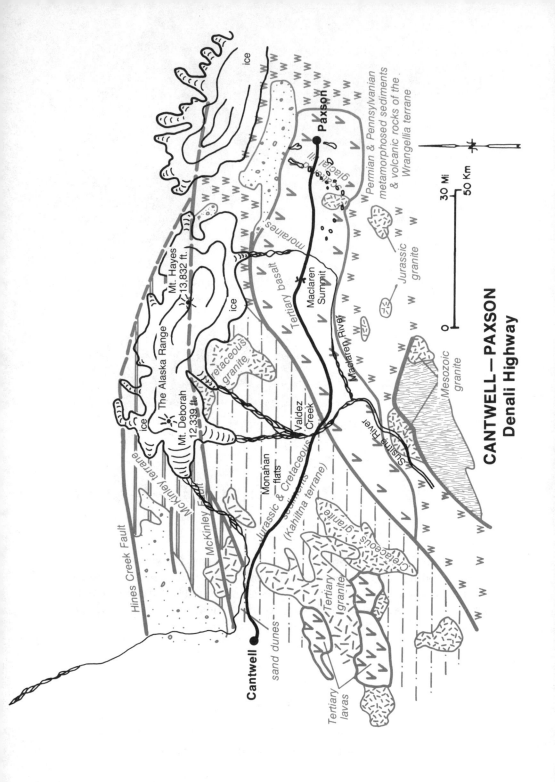

CANTWELL—PAXSON
Denali Highway

Hines Creek Fault

McKinley terrane

ice

The Alaska Range

Mt. Hayes
13,832 ft.

Mt. Deborah
12,339 ft.

ice

McKinley Fault

Jurassic & Cretaceous
sediments
(Kahiltna terrane)

Cretaceous
granite

Monahan
flats

Valdez
Creek

sand dunes

Tertiary
lavas

Tertiary
granite

Cretaceous granite

Tertiary basalt

Maclaren Summit

Maclaren River

Susitna River

Mesozoic granite

moraines

glacial till

ice

Paxson

Permian & Pennsylvanian
metamorphosed sediments
& volcanic rocks of the
Wrangellia terrane

Jurassic
granite

Cantwell

30 Mi

50 Km

rivers, which the wind can pick up and redeposit in dunes. Gulkana and Nenana glaciers are visible north of Monahan Flats where they flow down from the Alaska range. At least some of the silt in the dunes along the highway came from the floodplains of rivers draining these glaciers.

Bedrock

The first good roadside outcrop east of Cantwell exposes dark Jurassic and Cretaceous shales that contain coal. The sedimentary rocks were intruded by light-colored igneous dikes that contain over-sized crystals of plagioclase feldspar. A prominent small fault displays slickensides and breccia; the fault cuts through both the sedimentary rocks and the igneous intrusions. Shales, mudstones and sandstones account for nearly all the bedrock between Cantwell and the Susitna River.

Valdez Creek, which enters the Susitna River just east of the highway bridge, is one of the richest active placer streams in Alaska. Its headwaters are in metamorphosed sedimentary and volcanic rocks that were intruded by felsic and mafic magmas. The most valuable placers are in old channels buried beneath younger glacial deposits. A good assortment of heavy minerals appears in the pan concentrate.

Nearly all of the bedrock between the Susitna River and Paxson is greenstone or metamorphosed basalt, and other lavas. It erupted in early Tertiary time and lies on top of Wrangellia. The green colors are from minerals such as chlorite and actinolite that form when basalt is slightly metamorphosed.

Slickensides and breccia are apparent in this small fault that cuts Mesozoic sediments. —DO

205

Glacial Landscapes

The Maclaren River Valley is outstanding for its display of glacially deposited landscapes. The road crosses a moraine before dropping down into the valley, which is sprinkled with little kettle lakes. There is an especially good view from the west side of the valley.

In the Tangle Lakes area, the road goes through lowlands littered with glacial till. The unevenly distributed deposits are arranged in a knob and kettle topography. Kettle ponds formed when stagnant blocks of ice melted out of the till. A relatively impermeable cap of silt and clay that settled from a glacial lake covers the landscape and helps the ponds hold water.

Not long after the ice withdrew, about 10,000 years ago, people moved into the area. Stone-age hunters spent time on the knobs and ridges in the Tangle Lakes area, watching the lowlands for game. While they watched and waited they worked on their flint arrowheads; many fragments have been found on strategically located knobs. The artifacts include imperfect blades, points, and the cores of flint nodules from which chips were taken for the manufacture of spear points and knife blades. During the Ice Age, sabertooth tigers and woolly mammoths roamed the landscape and were hunted for their meat and hides. Until their decimation by modern hunters with rifles, large herds of caribou migrated through this area each year. Now only a small herd is left around Nelchina.

IV
NORTHERN ALASKA

Elliott Highway:
Fox—Livengood—
Manley Hot Springs
152 mi./245 km.

The Elliott Highway winds for 152 miles through gold mining country to Manley Hot Springs. East of Livengood, the road goes through ancient metamorphic rocks of the Yukon-Tanana terrane; to the west the rocks belong to the Manley terrane. Manley Hot Springs was established as a homestead in 1902; a four-story hotel was added in 1907. Until 1957 it was known as Baker Hot Springs.

Livengood

Gold was discovered on Livengood Creek in 1914 by Jay Livengood and N.R. Hudson. The creek was mined for about a year until the real paystreak was discovered in the benches above Livengood. By 1918, the population of Livengood reached 1500 people and the mines had four miles of interconnected tunnels. The boom lasted until 1922, when the town began to lose population. Only a handful of people remain.

Terranes

For about 30 miles north of Fairbanks, the Elliott Highway crosses the late Precambrian to early Paleozoic metamorphic rocks of the

The Chatanika ditch used a suspension bridge to cross the Chatanika River, 1927. —U.S. Geological Survey photo by P.S. Smith

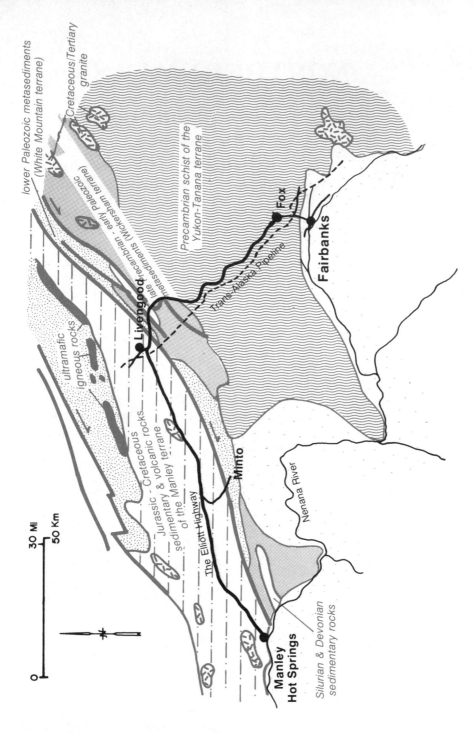

FOX—LIVENGOOD—MANLEY HOT SPRINGS
Elliott Highway

lower Paleozoic metasediments
(White Mountain terrane)

Cretaceous/Tertiary granite

Precambrian schist of the Yukon-Tanana terrane

early Paleozoic

late Precambrian (Wickersham terrane) metasediments

Livengood

Fox

Fairbanks

Trans-Alaska Pipeline

ultramafic igneous rocks

Jurassic - Cretaceous sedimentary & volcanic terrane of the Manley terrane

Cretaceous rocks

Minto

The Elliott Highway

Nenana River

Manley Hot Springs

Silurian & Devonian sedimentary rocks

30 Mi

50 Km

The hills north of Fairbanks are made of deeply weathered schist, here with a cap of windblown glacial silt or loess. VW for scale.

—DO

Yukon-Tanana terrane. These rocks include the Birch Creek schist, the source of a lot of gold in the Fairbanks mining district. North of Glove Creek, the road crosses northeast-trending belts of Paleozoic and Cretaceous sedimentary and volcanic rocks. The Manley terrane contains sedimentary rocks and lavas plastered against the Yukon-Tanana terrane and younger sediments formed by erosion of the colliding terranes. For better than half the way from Fox to Livengood the road travels across rounded hills of the Yukon-Tanana terrane. The rocks are so deeply weathered that in places you can dig through the rotten schist with a shovel. The same billion year-old rocks seen here are also in Fairbanks, Delta Junction, Tok, and extend to the east across the Yukon border.

North of milepost 68 the road cuts through deeply weathered serpentinized peridotite. The surfaces of the rocks show slickensides, striations produced as they were squeezed along faults like grease. Periodotites are even more rich in iron and magnesium minerals than mafic rocks; compositionally they are at the far end of the spectrum of igneous rocks. They are dark green, very dense, sometimes have a soapy feel, and weather to a rusty brown color in outcrop. Their native habitat is in the mantle.

From Livengood to Manley Hot Springs the road follows the Manley terrane, a collection of Mesozoic sedimentary and volcanic rocks that were strongly deformed and intruded by Cretaceous granite. Manley Hot Springs is on one of the granite plutons.

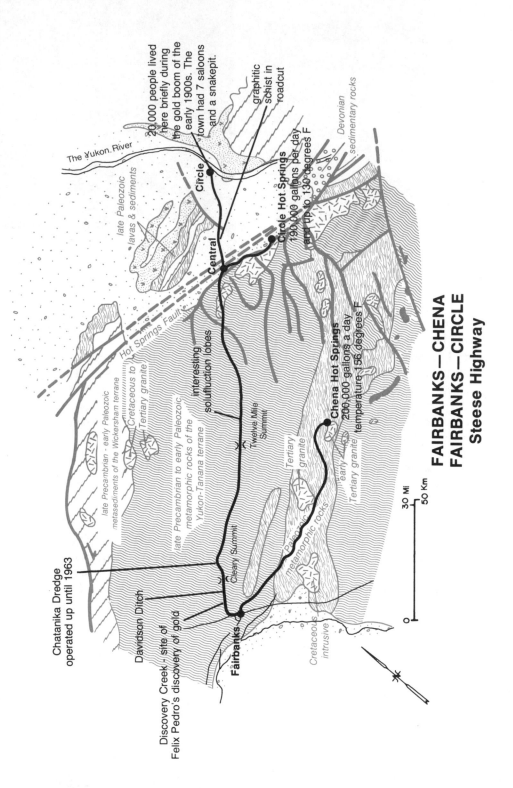

FAIRBANKS—CHENA
FAIRBANKS—CIRCLE
Steese Highway

30 Mi
50 Km

Chatanika Dredge
operated up until 1963

Davidson Ditch

Discovery Creek - site of
Felix Pedro's discovery of gold

The Yukon River

20,000 people lived
here briefly during
the gold boom of the
early 1900s. The
town had 7 saloons
and a snakepit.

graphitic
schist in
roadcut

Circle

Central

Circle Hot Springs
190,000 gallons per day
and up to 130 degrees F

Devonian
sedimentary rocks

late Paleozoic
lavas & sediments

Hot Springs Fault

late Precambrian - early Paleozoic
metasediments of the Wickersham terrane

Cretaceous to
Tertiary granite

interesting
solifluction lobes

late Precambrian to early Paleozoic
metamorphic rocks of the
Yukon-Tanana terrane

Twelve Mile
Summit

Tertiary
granite

Chena Hot Springs
200,000 gallons a day
early Tertiary granite temperature 156 degrees F

Cleary Summit

Paleozoic
metamorphic rocks

Fairbanks

Cretaceous
intrusive

*An ice wedge
exposed by placer
miners, 1949.*
–U.S. Geological Survey
photo by O.J. Ferrians

STEESE HIGHWAY

Fairbanks—Chena
Fairbanks—Circle
57 mi./91 km.
162 mi./261 km.

Roads to Circle and Chena Hot Springs wind their way through rounded hills of weathered schist, each leading to a hot springs issuing from granite. The roads go through gold mining country where millions of ounces of gold have been recovered.

The Chatanika dredge operated until the early 1960s near Fairbanks. —DO

The Chatanika Dredge

A gold dredge along the Steese Highway, across the road from the Chatanika Lodge, has sat idly since 1963. At one time many of these monsters operated in the Fairbanks region. After the richest placers had been worked out by hand in the early part of the century, large blocks of claims were bought up by the Fairbanks Exploration Company. The company had as many as eight dredges operating at one time, recovering gold from low-grade and previously worked gravels.

A gold dredge floats in a pond which it has dug for itself. It scoops up gravel with a bucket chain, generally down to the bedrock. The inside of a dredge is a factory for gleaning gold from gravel; the gravel is sorted, sifted and washed. Gold is trapped in riffles and mats in a sluice. After being processed, the waste gravel is sent out on a conveyor and deposited on a tailings pile behind the dredge. So the machine works its way through the gravel by digging in front and dumping behind.

Mucking and Stripping

Around Fairbanks, the frozen gold-bearing gravels are buried beneath a layer of frozen loess. Miners refer to it as 'muck' because of its organic content and fetid smell. Before mining could begin, the muck had to be thawed and removed. The usual procedure was to wash away the muck a few inches at a time with 'giants,' large hydraulic nozzles.

After stripping the overlying muck, the frozen gold-bearing gravels had to be thawed before being worked by a dredge. This was done through a grid of pipes drilled into the ground. Stream water a few degrees above freezing was pumped into the ground at low pressure. The frozen gravels thawed at a rate of up to a foot per day. Only after the lengthy process of stripping and thawing could the dredges operate.

The Davidson Ditch

The Davidson ditch was constructed in the late 1920s to bring water from the Chatanika River to Fox for stripping, thawing and dredging Goldstream Valley. The 90 mile-long pipe bears the name of the chief engineer who dreamed up the project. The modern Steese Highway follows a wagon road built to support construction of the pipeline.

Hot Springs

The hot waters of Chena and Circle hot springs issue from granitic plutons that intruded the Birch Creek schist. Granitic plutons are favorable sites for development of hot springs because of the physical properties of granite and because water may circulate to great depth in fault zones.

The Davidson Ditch was constructed in the late 1920s to bring water from the Chatanika River to Fox. The ditch provided water for the stripping, thawing, and dredging of the Goldstream Valley. —DO

213

Hot springs seem to like granite, partly because it is hard and relatively insoluble. Fractures in granite are not easily closed or sealed. Hot water from Chena Hot Springs may have been circulating through fractures in granite to a depth of two miles, where the rock is hotter than the surface boiling point of water. Rocks everywhere get hotter with increasing depth.

Another reason hot springs seem to like granitic plutons is that they are relatively rich in radioactive isotopes of elements such as thorium, uranium, and potassium. When they decay, these elements release heat and, although they are only a very tiny fraction of the total amount of rock, their presence is enough to raise the temperature of the rock.

Typically, heat flow in granitic plutons may be 50 percent higher than surrounding areas. So, contrary to what you might think, Chena and Circle Hot Springs do not get their heat from molten rock beneath the surface. That may be the case with hot springs near volcanoes, but almost certainly not here.

PIPELINE HAUL ROAD
(DALTON HIGHWAY)

Livengood—Coldfoot
242 mi./390 km.

This may be the most mosquito infested segment of road in Alaska. In the summer it is either uncomfortably hot and the insects out of control, or it is raining. In the winter the temperature falls to seventy degrees below zero.

North of Livengood the Haul Road, or Dalton Highway, cuts through rounded hills of deeply weathered ancient rocks as it parallels the Trans-Alaska Pipeline. Along some stretches of road it's difficult to find natural outcrops, so the only rocks to look at are in the road-metal quarries. Between Livengood and the Yukon River, the rocks belong to the Tozitna terrane, a collection of gabbro and basalt, argillite, tuff, chert, dirty sandstone, conglomerate, and limestone. But these rocks aren't well exposed and a blanket of frozen loess covers any low bedrock outcrops. All exposed rock is deeply weathered. South of milepost 33, there is chert exposed by the side of the road that lies in fault contact with ultra-mafic rocks.

A few miles north of the Yukon River, the road crosses over to the Ruby terrane, a collection of schist and gneiss. The terrane is full of granitic plutons, and the road passes several. The granite has been reduced to coarse, sandy debris. After the mica and some of the feldspar grains weather, the grains of feldspar and quartz separate from the rock and pile up on the ground as coarse sand.

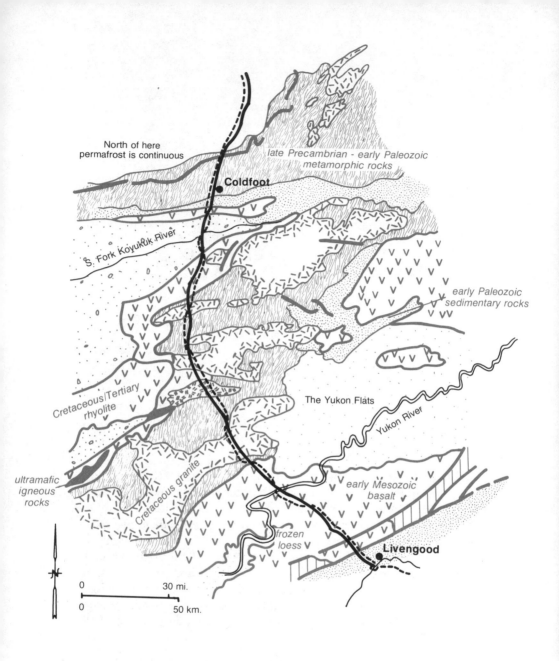

North of here
permafrost is continuous

late Precambrian - early Paleozoic
metamorphic rocks

Coldfoot

S. Fork Koyukuk River

early Paleozoic
sedimentary rocks

Cretaceous/Tertiary
rhyolite

The Yukon Flats

Yukon River

ultramafic
igneous
rocks

Cretaceous granite

early Mesozoic
basalt

frozen
loess

Livengood

0   30 mi.
0   50 km.

LIVENGOOD—COLDFOOT
Dalton Highway

Trans-Alaska Pipeline

Digging a trench and burying the Pipeline would certainly have been cheaper than raising it above ground, but raising the pipeline was necessary to avoid thawing the permafrost. Underground, hot oil would have melted ice around it and caused the ground to settle. More than half of the 800 mile length of the pipeline is elevated to avoid melting underground ice.

The stilts that support the pipeline are designed to refrigerate the ground rather than conduct heat into it. This is accomplished by using a liquid that has a very low boiling point and high vapor pressure. The liquid boils at a temperature below the freezing point of water. When heat is transferred to liquid in the bottom of the post, it boils and changes to gas. The gas then rises to the top, where it releases heat into the atmosphere through the radiator fins, then returns to the liquid phase. The fluid then drips to the bottom where it will rest until it absorbs enough heat to boil back to the top of the post.

The pipeline doesn't follow a straight course everywhere. The zig zag bends confer an accordian flexibility that allows the pipeline to expand and contract as the temperature changes, and to absorb earthquake waves without damages.

Permafrost

Discontinuous permafrost exists along the entire section of road, as in most parts of Alaska's interior. In many places the permafrost is a relic of the ice ages that are slowly thawing, creating thermokarst

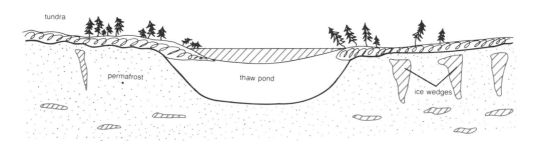

Cross section through permafrost region along haul road. Thaw ponds form as masses of ice within the soil melt, causing the ground to collapse. The drowning trees along the rim of the pond show that it formed quite recently.

features that form when the ground collapses as underground ice melts. The ice is thickest and most extensive in the fine-grained sediments along the Yukon River. Clay and silt can draw in water by capillary action. It then accumulates in ice masses that may exceed, by several hundred percent, the original mass of sediment. The pipeline is above the ground to prevent rapid thawing of the permafrost. If the ice beneath were to melt, the ground and pipeline might collapse.

A stunted black spruce forest is about all that manages to grow on the frozen loess; better drained areas underlain by gravel or bedrock support birch and other vegetation. In the summer the black spruce forest is boggy because surface water cannot drain through the permafrost. Ponds that form in the thawing permafrost are often rimmed by drunken spruce trees that fall over as the ground melts and collapses into the pond. These places are very difficult to build on; they are frightfully cold in the winter and in the summer the mosquitoes will literally eat you alive. No wonder hardly anyone lives here.

This mica schist has been deformed by Chevron folding. —DO

Coldfoot—Prudhoe Bay
242 mi./391 km.

Between Coldfoot and Prudhoe the Haul Road crosses the Brooks Range, and traverses a succession of rocks that range in age from late Precambrian to recent. People driving north encounter the oldest rocks first: the quartz-mica or chlorite schist, greenstone, phyllite, and quartzite between Coldfoot and Wiseman. These metamorphic rocks are more than half a billion years old. They are exposed in a broad band across the southern flank of the Brooks range.

The historic mining community of Wiseman lies about a mile west of the Haul Road. It sprang up in 1908 and reached its heyday in 1910 as a supply point for placer mining on Wiseman, Mascot and Nolan creeks, and the Hammond River. The town site was at the stopping point for supply boats going up the Koyukuk River. In the 1930s Bob Marshall used Wiseman as a jumping off point for his forays into the Brooks range. Only a handful of year-round residents live there today.

Between Wiseman and Galbraith Lake the bedrock is a series of sedimentary rock formations deposited in a shallow sea that persisted from Devonian through Permian time. The sea lay between

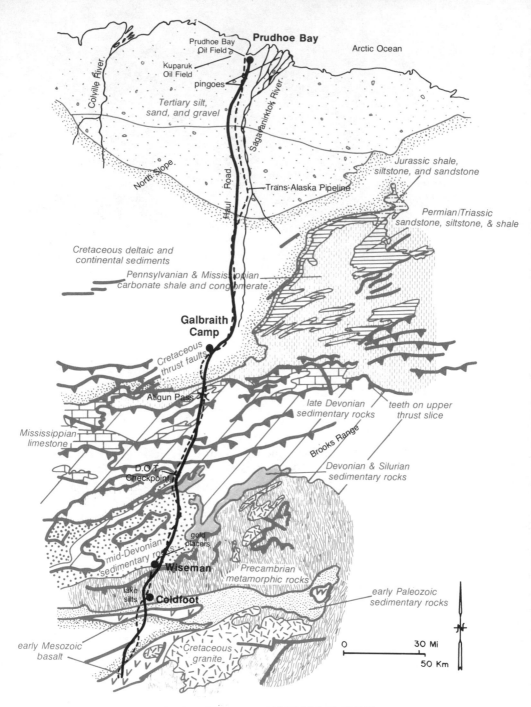

COLDFOOT—PRUDHOE BAY
Dalton Highway

Alaska and Siberia as North America and Eurasia embraced in a dance of continents. During late Paleozoic time, most of Alaska was a low plain with a few low hills. At times, sea level rose to inundate vast areas where coral reefs grew and sediments accumulated on the sea floors. During Mississippian time, as much as 2000 feet of carbonate sediments accumulated to form the Lisburne formation. Layers of sandstone, shale and carbonate accumulated through late Paleozoic time, then were pushed up in slices during Cretaceous time. Those thrust faulted slices now form the heart of the Brooks range. Many of the more prominent cliffs and peaks are made of limestone, a hard rock that resists weathering in the climate of the Brooks range.

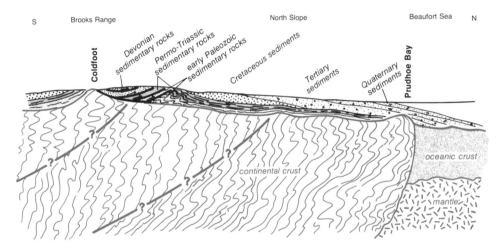

Cross section between Coldfoot and Prudhoe Bay. Thrust-faulted slices of Paleozoic rocks that core the Brooks Range give way to the gently dipping, younger sedimentary rocks of the North Slope.

North of Galbraith Lake, the road follows the Sagavanirktok River to the Beaufort Sea and crosses a series of sediments that were eroded from the Brooks range, then deposited on the North Slope. Cretaceous sedimentary rocks are exposed in a broad band that stretches from the Yukon border, to Cape Lisburne on the Bering Sea, and point Barrow on the Arctic Ocean. They are shale, sandstone, and gravel deposited originally in deltaic plains and alluvial fans from the erosion of the Brooks range. Younger sediments of Tertiary age are at the surface for the last 50 miles into Prudhoe Bay.

Pingoes

Pingoes are the only high ground on the North Slope of Alaska, an almost featureless flat plain that extends monotonously in all directions. They are giant pimples on the frozen tundra. Pingoes have an ice core that grows as underground water feeds it. In a milder climate that water would flow as a spring. In the cold arctic it freezes in the ground and expands with great force to form a blister on the surface, a pingo.

North Slope Oil

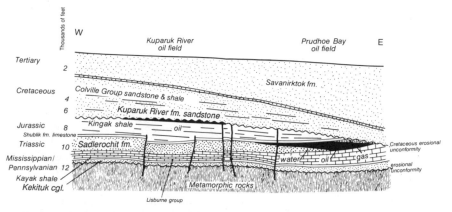

Cross section through North Slope oil fields. The organically-rich Kingak shale is the source of the oil which rises to reservoirs in porous sandstones.

The giant Prudhoe Bay oilfield contains almost 10 billion barrels of oil, more than any other oilfield in North America. The source of the oil is the dark, organically rich Kingak shale deposited during early Jurassic time. The exact chemical process that transforms organic material to oil is not completely understood. But petroleum, once formed, tends to rise because oil is lighter than the water that fills most sediments. Oil will float if it can find a path through the pore spaces and fractures.

The Prudhoe Bay reservoir is stratigraphically below, but at a higher elevation than the source rock from which the oil came. Paleozoic and Mesozoic sedimentary rocks dip slightly to the south in the Prudhoe Bay area and oil migrated up through the Sadlerochit formation until it encountered impermeable shale capping an unconformity, and could rise no farther. The Sadlerochit formation, composed of porous sandstone and conglomerate deposited during Triassic time, is the reservoir rock.

Since the discovery of oil at Prudhoe Bay, there have been several smaller strikes, most notably near the Kuparuk River west of Prudhoe Bay and along the barrier islands to the north. These new oil fields would not be economical to develop by themselves but because the pipeline and pumping stations are already built, they are being developed. The oil comes from the same source rock as that at Prudhoe Bay but has accumulated in other porous reservoir rocks.

The Kuparuk oil field is 10 to 30 miles west of the Prudhoe Bay oil field. It is estimated to hold more than four billion barrels of oil similar to that at Prudhoe Bay; about 1.5 billion barrels are recoverable with present technology. The oil is trapped in the Kuparuk River sands beneath the great Cretaceous unconformity that marks the top of most reservoirs on the North Slope.

Alaska's Dinosaurs

About 50 miles west of Prudhoe Bay along the Colville River, adult and young hadrosaur, tyrannosaur, and troodon dinosaur bones have been found in Late Cretaceous rocks. They record Alaska dinosaur life at the end of the Cretaceous when an asteroid impact is thought to have contributed to their worldwide extinction 65 million years ago. Alaskan dinosaur fossils have raised many questions about the ability of such large animals to survive at high latitudes during long dark seasons without hibernation, their possible migration patterns, and whether or not they were warm-blooded animals.

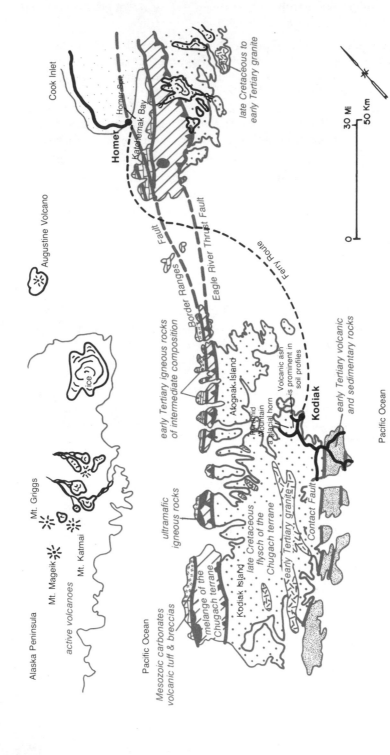

Alaska Peninsula

Mt. Griggs

Mt. Mageik

active volcanoes

Mt. Katmai

ice

Pacific Ocean

Augustine Volcano

Cook Inlet

Homer

Homer Spit

Kachemak Bay

Border Ranges Fault

Eagle River Thrust Fault

late Cretaceous to
early Tertiary granite

30 Mi

50 Km

Ferry Route

early Tertiary igneous rocks
of intermediate composition

Afognak Island

Pyramid
Mountain
a glacial horn

Volcanic ash
is prominent in
soil profiles

Kodiak

early Tertiary volcanic
and sedimentary rocks

Pacific Ocean

ultramafic
igneous rocks

late Cretaceous
flysch of the
Chugach terrane

early Tertiary granite

Contact Fault

Kodiak Island

melange of the
Chugach terrane

Mesozoic carbonates
volcanic tuff & breccias

Pacific Ocean

**HOMER—KODIAK
Ferry Route**

V
WESTERN ALASKA

Homer—Kodiak
110 mi./185 km.

Between Homer and Kodiak, air and ferry routes afford a view of fjords and cliffed islands. To the east, the heavily glaciated Kenai Mountains are visible on a clear day. To the west lie volcanic peaks of the Aleutian range.

Kodiak was the first Russian settlement in Alaska. In 1790, fifty-two Russians accompanied Alexander Baranof to Russia's new colony on Kodiak Island. For this outpost, they established a thriving fur trade that lasted until the near extinction of the sea otter. Their eighty years of residency left a mark that remains today.

Geologic Disasters

The town of Kodiak has been the victim of geologic disasters twice in this century.

On June 6, 1912, at 5 pm, coarse gray ash began to fall from the sky and it became as dark as night. The ash was coming from the then unknown Novarupta volcano, in what would later be called the Valley of Ten Thousand Smokes, from the west across Shelikof Strait. People could just barely breathe in the hot, nauseating atmosphere as they watched their homes collapse under the weight of the ash. To add to their fear, the ground shook terribly. By the time the ash stopped falling, it was 18 inches deep on the level and many feet deep in slides off the cliffs. It was months before the town got back to normal.

Rhythmic beds of shale and sandstone were deposited in deep water as part of the Chugach terrane.
—DO

Volcanic ash from the 1912 eruption of Novarupta is visible in the soil in many places around the island. The light-colored ash layer is toward the top of the soil profile, underneath the humus layer. It has the texture of fine-grained sand and is very abrasive.

On March 27, 1964, the ground began to rumble and sway in a rolling motion. The tremors increased and to everyone's terror, lasted for several minutes. Then, just when everyone thought that the worst was over, a giant swell appeared on the ocean and rolled toward Kodiak. The wave kept coming, lifting boats and then houses and cars as it swept through town. Wall after wall of water followed, destroying the wharf and the entire downtown. Along the coast, nearby villages and canneries suffered the same fate at the hands of the giant waves generated by the earthquake. The Good Friday earthquake of 1964, also known as the Anchorage earthquake, had visited Kodiak as well.

Chugach Terrane

Nearly all the bedrock exposed on the island is the dull, monotonous gray metamorphic rock of the Chugach terrane. Similar rocks of the same age exist in the Kenai Mountains, the Chugach Mountains, and Prince William Sound. These rocks were added as a unit to a growing Alaska during late Cretaceous time. They originated as a belt of sediments alongside a volcanic island arc somewhere to the south, and were carried north by drift of the Pacific plate, then plastered onto the continent.

Ice Ages

Pyramid Mountain, south of the airport, is a classic glacial horn that lies nestled in the hills behind the village of Kodiak. Notice that the surrounding hills are smooth to the level of the bottom of the horn, and rough above. That is the level to which the ice accumulated during the last ice age. Kodiak Island during the Pleistocene must have been a lot like the most heavily glaciated parts of the Kenai Mountains are today, with only a few nunataks poking through the ice.

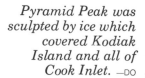

Pyramid Peak was sculpted by ice which covered Kodiak Island and all of Cook Inlet. —DO

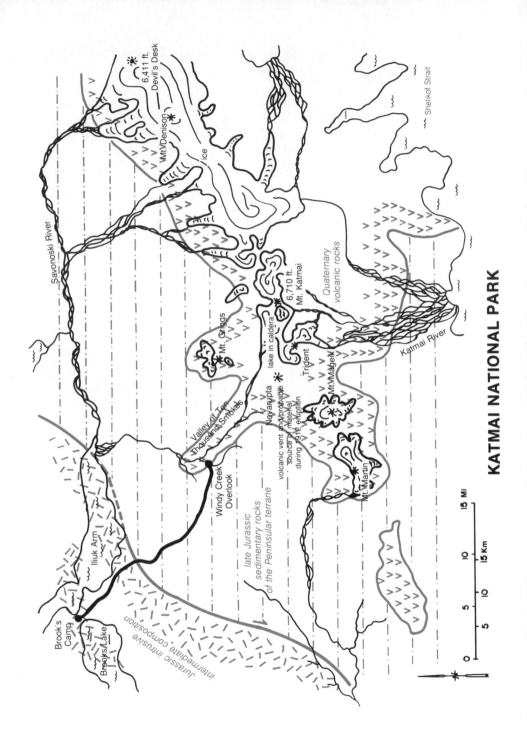

KATMAI NATIONAL PARK

Brook's Camp

Brooks Lake

Iliuk Arm

Savonoski River

Jurassic intrusive composition intermediate

late Jurassic sedimentary rocks of the Peninsular terrane

Windy Creek Overlook

Valley of Ten Thousand Smokes

Mt. Griggs

Novarupta

volcanic vent and probable source of material during 1912 eruption

6,411 ft. Devil's Desk

Mt. Denison

ice

lake in caldera

6,710 ft. Mt. Katmai

Trident

Mt. Mageik

Mt. Martin

Quaternary volcanic rocks

Katmai River

Shelikof Strait

0   5   10   15 MI

0   5   10   15 Km

228

Katmai visitors stand on tuff and pumice ejected by Novarupta's massive 1912 eruption.

—National Park Service
photo by R. Nichols

KATMAI NATIONAL PARK
1912 Eruption

Before the 1912 eruption, there was no Valley of Ten Thousand Smokes. There was a lush green valley with a mature forest growing in it. The valley was usually uninhabited, but once in a while people walked through on their way between Lake Brooks and the coast. The volcano that overlooked it had been known to rumble, but had never produced more than steam.

When Katmai awoke, it made its presence known to the whole world. Even without eyewitnesses, it was obvious from the beginning that this was one of the greatest eruptions in recorded history. Earthquakes that preceded the eruption shook villages along the coast as far as 130 miles away. Towns a hundred miles away were buried under a foot of ash and the noise of the eruption was heard at a distance of 750 miles. The eruption threw so much dust into the upper atmosphere that the warmth of the sun was diminished for many months afterward throughout the northern hemisphere.

Valley of Ten Thousand Smokes

The Valley of Ten Thousand Smokes was named by members of the 1916 National Geographic Society expedition, the first people to explore the volcanic wasteland. Hundreds of fumaroles steamed and hissed as water vapor escaped from the hot volcanic ash. Some of the fumaroles were so hot that if a piece of wood were held in the steam for a few seconds, it would immediately burst into flames when it was pulled out. Mount Griggs later was named to honor the leader of the expedition, the man who first described the Valley of Ten Thousand Smokes.

From the evidence it now appears that a glowing cloud of the type that a few years earlier destroyed St. Pierre in the West Indies swept down the valley, wiping out everything in its path. A glowing cloud is

The lower part of the Valley of Ten Thousand Smokes, pre-1920.
—U.S. Geological Survey photo by W.R. Smith

The village of Katmai was half buried under the ash of the 1912 eruption. —U.S. Geological Survey photo by G.C. Martin, 1914

a mixture of red hot ash and gases that sweeps down the slopes of a volcano. This one roared down the valley for seventeen miles, then spread out like a flood into the adjoining drainages. Only near its edges can so much as charcoal remnants be found of the mature forest that existed before the eruption. Because the old topography is not known, the total amount of material erupted is open to conjecture. It must have been on the order of several cubic miles, enough to fill a train stretching several times around the globe. The material cooled into an easily erodible tuff in which streams are presently carving badland topography. The glowing cloud was followed almost immediately by eruptions that buried everything in more layers of ash and pumice.

The source of the volcanic ejecta was not the summit of Mount Katmai, as you might guess, but a vent at the base of the volcano. Today a smaller and younger volcanic cone, Novarupta, lies over the vent. The rhyodacite plug that forms the center of Novarupta is surrounded by blocks erupted before extrusion of the plug, the last act of the waning eruption.

The caldera on the summit of Mount Katmai formed as the top of the mountain collapsed after the magma chamber beneath was emptied through the vent at Novarupta. The summit crater and side vent were connected to the same plumbing system and the magma, which exerted tremendous steam pressure, blew out near the base of the mountain and caused the collapse of the summit. The crater lake inside the summit is one of the most awesome vistas in the world; the color of the water was first described by Griggs as a "vitriolic, metallic, robin's egg blue." The unique color comes from glacial silt and sulfur in the water. Volcanic heat keeps the surface of the lake free of ice most of the year. The two glaciers that descend into the lake could not have existed before the eruption; they are unique among glaciers because we know almost exactly how old they are.

Alaska Peninsula Volcanoes

Mount Katmai on the Alaska Peninsula is part of a volcanic chain, a string of nearly a hundred volcanoes extending out to the Aleutian Islands and created by the subduction of the Pacific plate beneath Alaska. Several of Katmai's neighbors are active and likely to erupt again in the near future. Mount Mageik and Mount Trident have each erupted several times since people started keeping records after the 1912 eruption of Novarupta. Mount Griggs and Mount Martin often spout steam and look as if they could erupt at any time. Just about any volcano that has a classic cone shape can be considered active because erosion alters the shape of inactive volcanoes in a relatively short time.

The volcanoes rest on older sedimentary Peninsular terrane rocks that were plastered onto the North American continent during late Jurassic and Cretaceous time. These marine shales, siltstones, and sandstones are many thousands of feet thick. The Jurassic batholith now exposed along Iliuk Lake and Lake Grosvenor formed early on during the accretion process when sediments were dragged down in a subduction zone to a depth where the temperature and pressure allowed partial melting to form granite magma. The present episode of volcanism has been going on intermittently since perhaps Eocene time and is evidence that subduction is continuing. Twenty million years from now, granitic rock at the root of Mount Katmai and its neighbors may be exposed when the volcanic and sedimentary rocks are eroded away. At that time only remnants of volcanic rock may survive in protected places to serve as evidence of the eruptions that happened here.

A team of researchers returned to Mount Katmai in the 1990s to study the nature of such explosive volcanoes. The eruption of Mt. Pinatubo in the Philippines on June 12, 1991 was successfully predicted, in part, because volcanologists recognized that its ancient accumulations of violently erupted volcanic material closely resembled those found in the Valley of Ten Thousand Smokes.

NOME

The Gold Rush

Nome sprang up practically overnight when word leaked out in 1898 of the discovery of gold on Anvil Creek by the Three Lucky Swedes. By the next summer there were 10,000 people in the area, twice that the following year when gold was discovered in the beach sands. Wild and lawless, Nome was suddenly the largest city in Alaska. People lived in tents until lumber arrived by ship. With no civil authority, thievery and claim jumping were rampant. The army was sent in to maintain law and order.

Some of the placers were fabulously rich. The Caribou Bill claim, on a headwater tributary of Anvil Gulch, had gravels that contained nearly three pounds of gold in a cubic yard. Overnight, sourdoughs became millionaires. Saloons in Nome sought their patronage by offering only the finest of whiskeys and cigars. After gold was discovered in the beach sands, there was enough for everyone. A tent city rose on the beach. Individual miners controlled the ground only as far as a shovel would reach because of a legal ruling that claims could not legally be staked on the beach. A sourdough could accumulate twenty to a hundred dollars a day working the beach sands with hand tools.

Eventually many of the richest paystreaks played out and mining companies bought up groups of claims and dredged large amounts of lower paying sand and gravel. Today, the remains of as many as forty old dredges lie rusting around Nome. For those arriving in Nome by airplane, there is an old dredge 1/4 mile from the airport. An active dredge operates one-half mile west of town, but visitors are discouraged.

Gold Placers

All placers begin with the weathering of rocks that contain heavy and resistant minerals. A placer is formed by mechanical concentration of the heavy minerals after their release from the parent rock. The final size and richness of a placer depend more upon an abundant supply of parent material and favorable conditions for concentration than on the richness of the lode source. There are many small gold-bearing veins in the hills around Nome, but nearly all the gold has

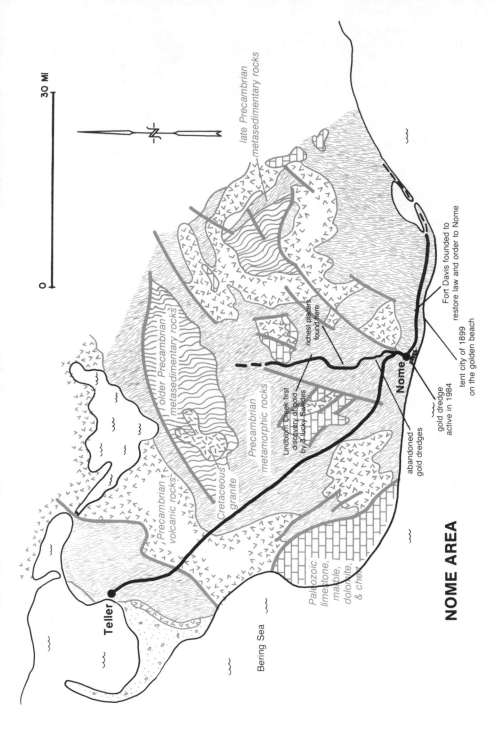

NOME AREA

late Precambrian metasedimentary rocks

older Precambrian metasedimentary rocks

Precambrian metamorphic rocks

Cretaceous granite

Precambrian? volcanic rocks

Paleozoic limestone, marble, dolomite, & chert

Bering Sea

Teller

Nome

Lindblom Creek: first discovery of gold by 3 lucky Swedes

richest placers found here

Fort Davis founded to restore law and order to Nome

tent city of 1899 on the golden beach

gold dredge active in 1984

abandoned gold dredges

30 MI

0

1901 Nome during the gold rush days. —U.S. Geological Survey photo by W.C. Mendenhall

come from stream and beach placers. The gold in stream placers around Nome came from lode deposits in the hills and from benches where some of it had been previously concentrated. Reconcentration created even richer placers, some of the richest in the world.

Gold nuggets can sometimes be found where a stream runs over bedrock, or on top of bedrock that has been buried by stream sediment. Concentration occurs during repeated scour and fill, events which may occur only rarely, during floods. A good place to look for nuggets is in holes in the bedrock or in cracks where the rock layering is nearly vertical. The dense gold nuggets lag behind in the cracks while sand and gravel move downstream.

In the gold rush days, miners on the beach kept an eye open for "ruby" sands—thin layers of garnet-rich sand. Both gold and garnet are dense minerals that are concentrated by the action of the waves. Today, ruby sands are hard to find and their place has been taken by analogous concentrations of iron particles, as the waves and sand grind away at the rusting hulks of gold rush mining equipment, and then concentrate the particles as they did the garnets and the gold.

Gold in the beaches almost certainly came down the streams from the hills north of Nome, and must have spent time in placers along the way. By the time the gold had made its way to the ocean it had been ground to fine dust.

Several ancient beaches, as well as the one along the present waterfront, contain gold placers. The beaches were stranded or submerged by changes in sea level. Mining companies are waiting for a rise in the price of gold before they begin their planned offshore mining of submerged beach placers.

Panning the beach for gold near Nome about 1900.
—U.S. Geological Survey photo by W.R. Smith

Discovery Claim is about a mile north of the Department of Transportation information sign of Anvil Creek north of Nome. This is where the Three Lucky Swedes discovered gold in 1898 and became rich men. The stream cobbles in Anvil Creek are metamorphic rock of the Seward terrane, mainly late Precambrian and early Paleozoic schist. Some are close to a billion years old. They are at the tail end of a long line of similarily old rocks that runs along the spine of the Rocky Mountains and into Alaska, but their origin is a mystery. These rocks may have originated somewhere in the western Pacific, and later added to the western end of the North American continent.

Bering Sea

The shallow Bering Sea was dry land for long periods of time during the Pleistocene glacial epochs. If scientists are correct in estimating a maximum 180 meters (550 feet) of sea level lowering during ice-age maxima, then at times there must have been dry land from Point Hope south to the Aleutians and from Siberia east to Alaska. The Bering Land Bridge was dry long enough for forests and tundra to become established and at times bore no resemblance to dried up ocean bottom. Great rivers developed in this vast area; their channels have been identified on the bottom of what is now the Bering Sea.

During the ice ages, the Bering Land Bridge was a migration route for humans and other species who made their way to North America from Asia. Ancestors of native Americans made the trip to the New

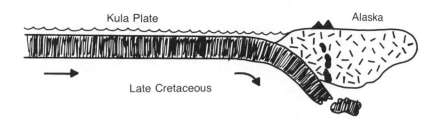

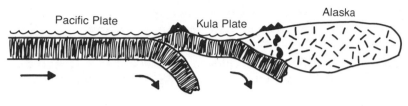

Late Tertiary

Evolution of the Bering Sea. Late Cretaceous subduction of the Kula Plate ceased when the plate broke and jammed the trench. Subduction shifted west to the Pacific Plate, stranding the Kula Plate under a shallow sea.

World by this route many thousands of years ago. Musk oxen were driven to extinction in Alaska during this century, but thanks to the land bridge, the animals had relatives in Siberia that could be the breeding stock for a new generation. They again range all across northern Siberia and Alaska.

The Bering Shelf is a leftover from the Kula plate which broke into pieces, jamming subduction on its northern side and beginning subduction to the south. The part of the Kula plate that was stranded is the only part that remains now; the rest of the plate has disappeared beneath Alaska. Subduction continues along its southern margin as the Pacific plate is diving under the remnant of the old Kula plate.

Glossary

Amygdule — a gas escape hole left in lava that has been later filled with minerals such as quartz or calcite.

Amphibolite — a type of rock formed by regional metamorphism of basalt under medium to high pressures and temperatures.

Andesite — a dark-colored, fine-grained igneous rock that is the extrusive form of diorite. It is between basalt and rhyolite in chemical composition and was named after the Andes Mountains in South America where it is common.

Argillite — a weakly metamorphosed mudstone or shale that is used to make beautifully carved bowls, totem poles and other artwork in southeastern Alaska.

Arkose — a sandstone made up mostly of quartz with at least 25 percent feldspar minerals. Arkoses form commonly in continental areas near young granitic mountain chains.

Ash — tiny fragments of volcanic rock blown into the air by escaping steam and distributed across the countryside near an erupting volcano.

Basalt — a dark-colored igneous rock that is made up mostly of the minerals plagioclase feldspar and pyroxene. It is the very abundant extrusive equivalent of gabbro.

Carbonates — rocks such as limestone or dolomite that are formed by the combination of atoms of calcium or magnesium with carbon and oxygen.

Clinopyroxene — a type of pyroxene mineral containing lots of calcium and usually magnesium.

Conglomerate — a sedimentary rock composed of rounded pebbles, cobbles, or boulders in sand. A gravel rock.

Diorite — an igneous rock that forms by slow cooling beneath the earth's surface and is between gabbro and granite in chemical and mineral composition. It is made up of feldspars, hornblende and pyroxene with little or no quartz.

Flysch — a collection of marine sediments shed by a risng mountain chain as it is uplifted and eroded. The sediments are mainly silt and sand.

Formation — The basic subdivision of sedimentary rocks that can be mapped from place to place based on descriptive characteristics.

Gabbro — Dark-colored intrusive igneous rock that is composed of the iron- and magnesium-rich minerals augite and sometimes olivine, also with calcium-rich plagioclase feldspars. When the same magma erupts at the surface, it forms basalt.

Garnet — a dark red semi-precious mineral common in metamorphic rocks; its hardness makes it useful as an abrasive.

Gneiss — a regional metamorphic rock that has a striped appearance caused by light-colored bands of granular minerals alternating with darker bands of platy or flaky minerals.

Granite — a light-colored, coarse-grained igneous rock that is made up mostly of plagioclase, potassium feldspar, and quartz; but which may also contain a little mica or hornblende.

Granodiorite — an intrusive igneous rock with a chemical composition between diorite and granite. It is made up of quartz, plagioclase, and a little potassium feldspar, biotite and hornblende.

Greenschist — a metamorphic schist that has a greenish color due to the presence of the minerals chlorite, epidote, or actinolite.

Greenstone — volcanic rocks that have been metamorphosed and have grown green metamorphic minerals. Not as platy as a schist.

Greywacke — a dirty sandstone made up of coarse grains of quartz, feldspar, and volcanic rock. This sandstone forms as a turbidite deposit on the sea floor, and the "dirt" forms later with chlorite alteration by circulating sea water.

Group — two or more formations that occur together.

Hornblende — the most common mineral of the amphibole group usually occurring in elongate dark crystals.

Loess — silt picked up from glacial streambeds and redeposited by the wind.

Magnetite — a black strongly magnetic iron mineral that is an important ore of iron and occurs commonly as a heavy mineral in sands.

Melange — a mixture, on a grand scale, of crunched rocks that have been caught between colliding plates.

Moraine — a mound or ridge of an unsorted mixture of silt, sand, and gravel (glacial till) left by a melting glacier.

Ophiolite — a collection of rocks formed on the sea floor.

Olivine — a glassy green or brown mineral that is common in gabbro and basalt. Large crystals of olivine occur in basalts from Hawaii and even make beaches of green sand there.

Outwash — sand and gravel washed by streams from the front of a glacier.

Pelites — sedimentary rocks made of mud.

Phyllite — a metamorphosed fine-grained sedimentary rock that has been cooked more than slate but not as much as schist.

Pluton — a body of intrusive igneous rock, commonly granite or granodiorite.

Pyroxenite — an ultramafic igneous rock composed mainly of pyroxene.

Quartz — composed entirely of silica, it is one of the most common rock-forming minerals.

Rhyolite — a group of extrusive igneous rocks that have the same chemical make up as granite.

Roof pendant — an isolated mass of the overlying cap rock that is surrounded by an intrusive body.

Schist — a common metamorphic rock with platy or aligned minerals.

Skarn — rocks composed mostly of calcium silicate minerals such as pyroxene, garnet, and epidote; originally limestones or dolomites that were later altered by the addition of silica from a nearby granite.

Strike-Slip Fault — a fault showing sideways movement or offset along a near-vertical fault.

Tectonic — the force responsible for a large-scale deformation of the earth's crust.

Terrane — fragments of a continent, ocean crust, volcanic arc, or disrupted combinations of these that share a common geologic history and have often travelled far from where they originated.

Thrust Fault — a fault dipping less than 45 degrees that formed by forces of horizontal compression. Generally the rock above the fault has moved upward and over the rock below it.

Till — unlayered and unsorted mixture of clay, silt, sand, gravel, and boulders that are deposited directly by a glacier to form a moraine.

Trachyte — a fine-grained extrusive igneous rock that is made up mostly of potassium feldspar with small amounts of dark minerals.

Transform Fault — a special type of strike-slip fault along a mid-ocean ridge or plate boundary.

Trench — a long, narrow, depression of the deep-sea floor that lies parallel to the continent. It may be a mile deeper than the surrounding ocean floor and may be thousands of miles long. It forms where oceanic crust is subducted or dragged down under continental or other oceanic crust.

Truncate — to cut off or shorten such as a fault might truncate a sequence of sedimentary rocks.

Turbidite — sands and muds that settle on the sea floor out of clouds of sandy, muddy water stirred up by turbidity currents. They form layers that grade upwards from coarse to fine particles and show moderate sorting of the particle sizes.

Ultramafic — an igneous rock composed entirely of dark minerals rich in magnesium and iron.

Varves — very thin sedimentary layers deposited in bodies of still water such as lakes. A dark and light pair of layers is generally thought to represent a summer and winter season in the lake. Each pair is then counted as one year of time.

Volcanoclastic — a sedimentary rock composed of broken fragments of volcanic rocks that have been transported from their source and cemented together.

Suggested Reading

Alaska Geographic, 1982, *Alaska's Oil/Gas & Minerals Industry*, Alaska Geographic Society, Vol. 9, No. 4.

Brooks, A.H., 1973 (reprinted), *Blazing Alaska's Trails*, University Alaska Press, Fairbanks, Alaska.

Cobb, Edward H., 1973, *Placer Deposits of Alaska*, U.S. Geological Survey Bulletin 1374.

Gehrels, G.E. and Saleeby, J.B., 1987, *Geologic framework, tectonic evolution, and displacement history of the Alexander terrane*, Tectonics, Vol.6, pages 151-173.

Hamilton, T.D., Reed, K.M., and Thorson, R.L., (eds.), 1986, *Glaciation in Alaska*, Alaska Geological Society.

Imlay, Ralph W., 1975, *Stratigraphic Distribution and Zonation of Jurassic (Callovian) Ammonites in Southern Alaska*, U.S.G.S. Prof Paper 836.

Jones, D.L., Cox, A., Coney, P. and Beck, M., 1982, *The Growth of Western North America*, Scientific America, Vol. 247, No. 5, pp 70-84.

Muir, John, 1979 (reprinted from 1915 edition), *Travels in Alaska*, Houghton Mifflin Company.

Orth, D.J., 1967, *Dictionary of Alaska Place Names*, U.S. Geological Survey Professional Paper 567.

Pewe, T.L., (ed.), 1965, *Guidebook to the Quaternary Geology of Central and South-Central Alaska*, reprinted by Alaska Division of Geological and Geophysical Surveys.

Pewe, T.L., 1975, *Quaternary Geology of Alaska*, U.S. Geological Survey Professional Paper 835.

Sabina, A.P., 1973, *Rocks and Minerals for the Collector, The Alaska Highway; Dawson Creek, British Columbia to Yukon/Alaska Border*, Geological Survey of Canada Paper 72-32.

Stone, David and Brenda, 1983, *Hard Rock Gold*, Vanguard Press, Inc., Seattle, Washington.

Tempelman-Kluit, D., 1980, *Evolution of Physiography and drainage in southern Yukon*, Canadian Journal of Earth Science, Vol. 17, pp 1189-1203.

United States Geological Survey, 1976-1986, *The United States Geological Survey in Alaska: Accomplishments During* (each of the years between 1976 and 1986), U.S.G.S. Circulars 751-B, 772-B, 804-B, 823-B, 844, 868, 939, 945, and 998.

Van der Voo, R., Jones, M., Gromme, C.S., Eberlein, G.D., and Churkin, M., Jr., 1980, *Paleozoic Paleomagnetism and Northward drift of the Alexander terrane, southeastern Alaska*, Journal of Geophysical Research, Vol. 85, No. B-10, pages 5281-5292.

Wolfe, Jack A., 1966 *Tertiary Plants from the Cook Inlet Region, Alaska*, U.S.G.S. Prof Paper 398-B.

Index

Check for our books at your local bookstore. Most stores will be happy to order any which they do not stock. We encourage you to patronize your local bookstore. Or order directly from us, either by mail, using the enclosed order form or our toll-free number, 1-800-234-5308, and putting your order on your Mastercard or Visa charge card. We will gladly send you a complete catalog upon request.

Some other geology titles of interest:

____ROADSIDE GEOLOGY OF ALASKA	15.00
____ROADSIDE GEOLOGY OF ARIZONA	15.00
____ROADSIDE GEOLOGY OF COLORADO	15.00
____ROADSIDE GEOLOGY OF IDAHO	15.00
____ROADSIDE GEOLOGY OF LOUISIANA	15.00
____ROADSIDE GEOLOGY OF MONTANA	15.00
____ROADSIDE GEOLOGY OF NEW MEXICO	14.00
____ROADSIDE GEOLOGY OF NEW YORK	15.00
____ROADSIDE GEOLOGY OF NORTHERN CALIFORNIA	15.00
____ROADSIDE GEOLOGY OF OREGON	14.00
____ROADSIDE GEOLOGY OF PENNSYLVANIA	15.00
____ROADSIDE GEOLOGY OF TEXAS	16.00
____ROADSIDE GEOLOGY OF UTAH	15.00
____ROADSIDE GEOLOGY OF VERMONT & NEW HAMPSHIRE	10.00
____ROADSIDE GEOLOGY OF VIRGINIA	12.00
____ROADSIDE GEOLOGY OF WASHINGTON	15.00
____ROADSIDE GEOLOGY OF WYOMING	15.00
____ROADSIDE GEOLOGY OF THE YELLOWSTONE COUNTRY	12.00

Please include $3.00 per order to cover postage and handling.

Please send the books marked above. I enclosed $_____

Name_____

Address_____

City_____State_____Zip_____

☐ Payment enclosed (check or money order in U.S. funds)

Bill my: ☐ VISA ☐ MasterCard Expiration Date:_____

Card No._____

Signature_____

MOUNTAIN PRESS PUBLISHING COMPANY
P.O. Box 2399 • Missoula, MT 59806
Order Toll-Free 1-800-234-5308
Have your MasterCard or Visa ready.